X = *t*

The Art of X-Ray Photography

N5B-1
T6·4
62·9·

4 - B

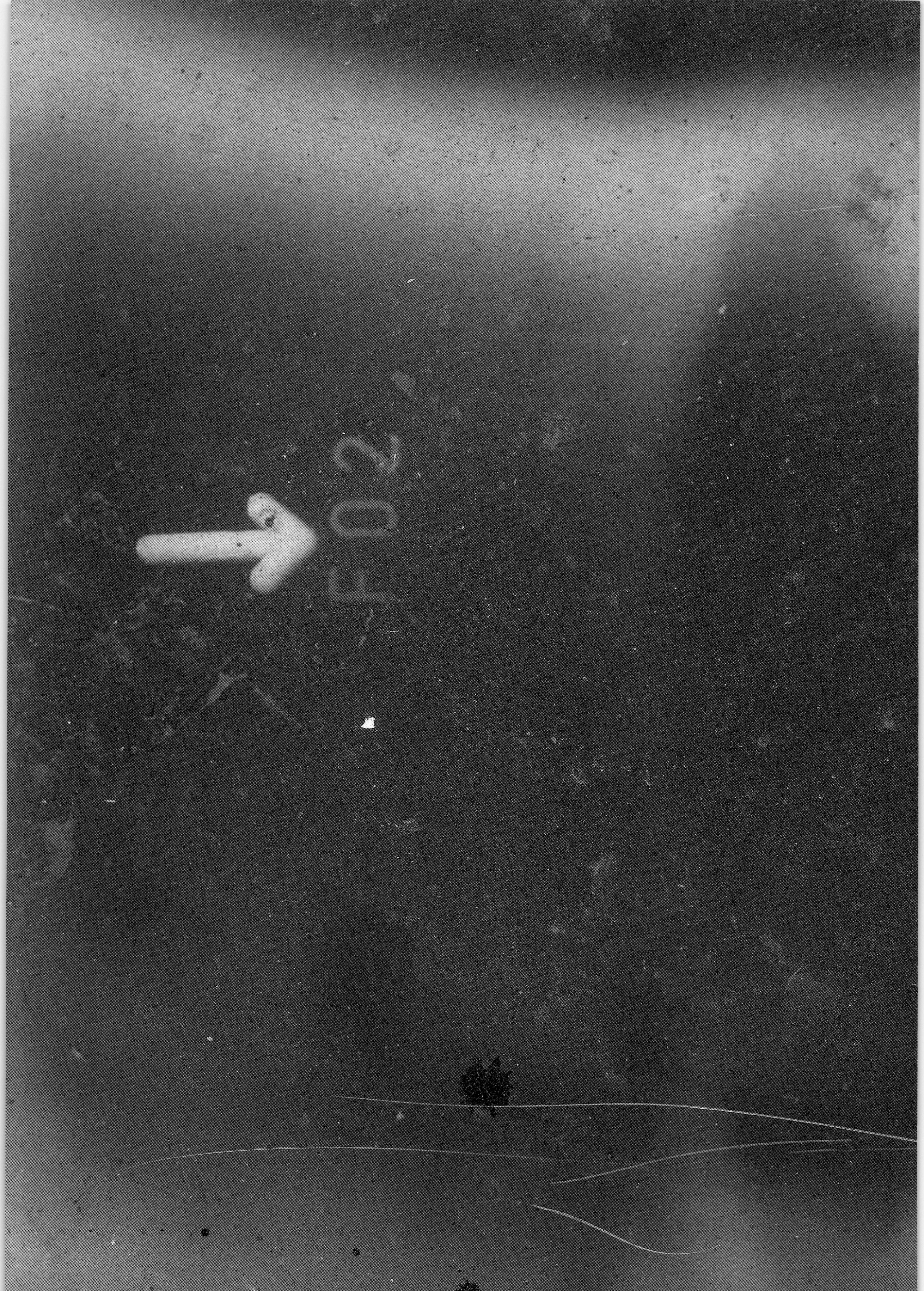
F02

43 3-A-1

$X = t$

The Art of X-Ray Photography

SEIJU TODA

Introduction by Max Kozloff

Hudson Hills Press
New York

First Edition

Published in the United States by Hudson Hills Press, Inc., Suite 1308, 230 Fifth Avenue, New York, NY 10001-7704.

Distributed in the United States, its territories and possessions, Canada, Mexico, and Central and South America by National Book Network.
Distributed in the United Kingdom and Eire by Art Books International Ltd.
Exclusive representation in Asia by EM International.

Editor and Publisher: Paul Anbinder

Copy Editor: Fronia W. Simpson

Art Director: Seiju Toda

Designer: Kumiko Shinozuka

Manufactured in Japan by Toppan Printing Company.

Agent: Jon Kramer

Library of Congress Cataloguing-in-Publication Data

Toda, Seiju, 1948–

X=t: the art of X-ray photography / by Seiju Toda : introduction by Max Kozloff.
p. cm.
Includes bibliographical references.
ISBN 1-55595-080-9
1. Radiography. 2. Radiography, Industrial. 3. Photography, Artistic. 4. Toda, Seiju, 1948– . I. Title. II. Title: Art of X-ray photography.
TR750. T63 1995
779'. 092—dc20 94-40107
CIP

Contents

X=*t*

The "X" stands for X-ray photography.

The "t" indicates time.

The "X" represents the hidden, the unknown.

The "t" is TODA himself.

Through the use of X rays,

"time" has become a visualized form.

Thanks to the use of X rays,

the mystery of beauty invisible to the eye,

is mystery no longer.

Art, for the first time, can be looked "clear through,"

at the hands of a scientific tool

and the equation, *X*=*t*.

The Concept

The photo series, *X=t*, is the world's first collection of X-ray photographs.

It is, in all probability, a form of art that no one has ever encountered before.

As a genre, photography has already attained its place in the world of fine arts, thanks to advanced camera technology and the refined aesthetic sense of photographers.

No one, however, has discerned the X ray's enormous potential for producing fine photography. It had remained something of a blind spot.

The X-ray camera's first and leading use was in the field of medicine.

It was used next in archaeology and in art restoration.

It has also become a valuable tool for checking the precision of industrial products.

It is, in fact, a highly developed, sophisticated tool.

It is, furthermore, capable of penetrating solid objects and photographing what is inside and invisible to the eye.

This supernatural power of the X-ray camera constituted the most important factor in the realization of this photo collection.

The birds, the animals, the flowers, the earth, the sky, and the human beings are in themselves all very beautiful on the surface.

An X-ray vision into their "inner space," however, affords an even lovelier view.

This is because the inside is where the essence of "life" resides.

And the concept that we call "time" is the key that preserves the beauty of "life."

Hence, we can say that "life" is also "time."

Today's architecture is a symbolic representation of this concept.

The exterior design of architecture effectively conceals the life that goes on inside.

Once the scale models were constructed, different kinds of creatures were placed inside and then photographed with an X-ray camera.

During photography, utmost attention was given to the effects of X-rays on "life."

Thus, the four dimensions of "life," "time," "architecture" (structure), and "the unknown" (the invisible), came to be imprinted on two-dimensional X-ray film, to create this remarkable photo-art collection.

X=t consists of 39 photographs, representing five years of work.

Metaphysics of the X Ray

Max Kozloff

As I begin this essay, I have in hand a ballpoint pen, made in Japan, bought in Spain. Its tip and push-button sections, along with its clip, lapped out like a reptile's tongue, are finished in matte aluminum. Its barrel is composed of grained wood, dyed gray. With its elegant blend of natural material and industrial surface, this implement is too distinctive to be a novelty item. Yet its manufacture is too sober and understated to pass for a luxury object. It impresses me, rather, as the product of a culture in which a restrained but sensuous craftsmanship—really an artistic consciousness—often inflects useful things.

The images by Seiju Toda, in this book, are processed with a like-minded care, but their white descriptions on black grounds, and their cryptic subjects, make them exotic presences. For they illuminate wooden "reliquaries," presented as the tombs of animal life, and they do so with the help of advanced technology.

In part, the sensibility that informs this work has a history that comprises Japanese decorative arts, lacquer boxes, tea paraphernalia, and sliding wall panels. The containers Toda builds from beautiful indigenous woods, to be pictured by X rays, retain some of the ceremonial aura which infuses the life of modest tools. Such objects had in Japan a calm, ritual significance to their users. Born in 1948, and therefore now in mid career, Toda came of age when it was still possible to assert a strong, if nostalgic, loyalty to this tradition. In 1974 he did a show called *Japanese Craftsman's Series*, whose subjects included the carpenter's plank and the plasterer's tile, photographs of which he photocopied, then silkscreened the results. In 1981, fascinated by textures, he pasted rice paper on canvas, then coated the surface with silicone, glass fiber, and oil paint. The added texturing was done for its own sake, though never so insistently as to obscure the original reference. Here is a mind with a taste for gradated transformations. Some of them are realized through overlays of transparent or translucent substance, manually applied; others, through graphic media, such as prints and photography. One is often asked to compare a version of an image that has been reproduced by ink processes, with another that is developed through optical-chemical means. Eventually, a viewer acquainted with Toda's past art notices that his desire to gratify the sense of touch is fused with excitement about the energy of a light that penetrates. With the extravagant reasoning of an artist, Toda has imagined sunbathing in a darkened room. This fantasy of 1969 unknowingly anticipates the work we see now, where invisible light waves travel through opaque surfaces and dark voids. It is a project called *X=t*, dating from 1986 to 1991. In order to visualize these pictures, the artist employs a hypersensitive, specially adjusted X-ray device developed in Japan for analyzing airplane parts. However he honors tradition, he also reveals himself to be entrepreneurial, by his recourse to state-of-the-art engineering, which exemplifies the innovative drive of the Japanese.

To American eyes, Toda's subject of vipers, and birds' wings or chimpanzees' arms, is less familiar than his accent on graphic mediation. We have long been viewers of an artistic emphasis placed upon the role the media have in influencing our lives. More often than not, this emphasis is colored by a critical attitude. Or perhaps, more explicitly, American artists are ambivalent toward their popular culture, which they feed off, in a half-enthusiastic, half-resentful mood. These X-ray pictures by a Japanese artist belong to another world. Toda studies the degrees by which various reproductive media affect our notion of physical objects, almost as if in a *metaphysical* spirit. The cool light seems to emanate from presences interred in matter; they look irradiated and degravitized. The artist's original wooden sculptures act only as secular points of reference to realize Toda's sacramental ends in pictorial form. For the aim is to "see through" solids, whose refinement of surface construction hints of, but cannot in itself set forth, the mystery of an inner life.

When first considering his unusual procedures, I thought to align and compare them with practices we know. Nothing, for instance, is in itself strange about his use of wooden pieces as setups for the finished image. The scenography of movies is our most common example of fabricated ensembles that have no reason for existence, other than to be photographed. And of course, commercial photo imagery depends quite often on temporary stagecraft. In American art photography, something parallel but also more obscure had happened in the work of a number of conceptualists focused on the paradoxes of depicting three-dimensional objects in two-dimensional images (James Casebere, Zeke Berman, Robert Cummings). Each in his own way, they were determined to show that perception of the original setup, when taken in by a photograph, involved the viewer in a number of optical errors and spatial misreadings. But at least a record of the setups' external surfaces had been recorded. Even if the objects themselves were never shown in themselves, we had some physical notion of them.

Toda's X ray, in contrast, cancels that notion. The schematic remembrance we have of the physical objects he depicts is sustained in his chosen medium only in the form of a photographic negative. In his backlit celluloid screens (and all other print versions he makes of them), we no longer deal with the body of things, only their husk and the skeleton within. Where objects were superimposed upon each other, their outlines are confused in the thickened expansion of white tones. As no doubt intended, the effect is ghostly.

In the history of Western modernism, that effect is recalled by the photograms of László Moholy-Nagy and the rayograms of Man Ray. Both those artists achieved their white-on-black tonalities by a kind of stenciling process, the placement of things on sensitive paper darkened by exposure to light. Being the reverse of normal photography because they revealed only what obstructed the light, not what was illuminated by it, photograms belong to the abstract tradition. In the sense that he conferred shape on light forms by cutting them out of the field, Man Ray evoked a space in which, according to Tristan Tzara, "the continuity of life does not encumber itself with the materialism of facts."

Even if we grant him his antimaterialism, however, Man Ray still confronts us with the floating presence of cleavers, meshes, and watchsprings, common objects whose identity is asserted through their outlines. Moreover, the range at which we view them is close, for these rayograms are literally contact prints. By contrast, Toda's icons are *revealed* to us at some indeterminate depth within their fields. Filigree or feathery configurations are shown despite, not by virtue of, the mass that embodies them. For all their common syntax of reversed shadows and lights, this difference separates Toda from his modernist predecessors. And even when I compare his images with the only other use of X ray I know of in American art, his result is still distinctive.

The main image in Robert Rauschenberg's *Booster*, a six-foot-high lithograph of 1967, is a vertically segmented X ray of a male human body, very likely the artist's own. In this case, the frankly acknowledged medical context is an imaginative feint, which blocks personal revelation even as it gives the most intimate self-exposure. Physically, no artist has ever uncovered more of himself than in this work, since he is shorn not only of clothes but of skin. At the same time, one is obliged to take the spectral white-on-black glow of the X ray literally, as simply an apparition supplied by a mechanized process that operates by the transmission of energy through invisible wave lengths. The X ray is merely another medium, another way of visualizing traces, like stone or wood rubbings. Rauschenberg followed in the Dada tradition pioneered by Man Ray, and here he partakes of some of the same etherealized worldliness as in the rayograms. Both artists seem to say that, even when devoid of substance, the world and the human forms within it can still be displayed in the form of shadows, beyond which there is no further meaning.

Though implicitly aware of Dada, Seiju Toda's X rays assume that light and shadow can be used to explore other dimensions than those in Western prototypes. I do not imply any judgment about the magnitude or depth of one artistic project as contrasted with another, but rather that his cultural, historical, and personal position allows Toda to contribute a fresh outcome to a Western impulse. For one thing, he works on the basis of a symbolic premise, which he tends to superimpose upon a Dada literalism. For another, he replaces the absurdist and skeptical tonality of Dada with a reference to a Japanese sense of materials. Even as an (obviously) precocious high school student, in 1965, he dedicated a performance piece to John Cage, and his description of it enlightens us about the terms of his dialogue with the West that continues throughout his career:

"I ran two pieces of transparent plastic tubing onto the stage, one coming from the right, the other from the left. Offstage, I stationed volunteers. Their job was to drop marbles into the tubes. The marbles, which were covered in fluorescent paint, glowed as they rolled down the tubes, through a darkened space that was illuminated only by black light. When the marbles got to the center of the stage, they dropped from the ends of the tubes into two receptacles—one, a wooden, Japanese-style washtub, the other a Western-style bucket made out of tin. That was my way of addressing the meeting of East and West, of paying tribute to John Cage, and of exploring chance effects of light and sound."

The determined effects of sound contrast, produced by the random plunk and plink of dropped marbles, certainly recall—therefore look back upon—Cageian strategies. And one should not forget that Cage's involvement with Zen Buddhism might well have struck a young Japanese as a reflection of an element in his own culture. But the thought of those little, moving, fluorescent spheres in the dark also strikes me as a wonderful anticipation of Toda's X rays to come. For even from youth, it seems as if the artist's dramaturgy was often to be formulated by a kind of magical illuminism.

Toda studied architecture, and he admires Arata Isozaki and Tadao Ando. It may be too much to say that their influence can be felt in his constructions, but there can be little doubt of his emphasis on the tectonic integrity of his wooden pieces. They are objects placed on pedestals, and they often have shafts, "floors," wooden girders, doors, and buttresses. However varied, their function is to encase stuffed or possibly dried animals like stingfish or creature's parts, such as pigeon's wings. The fish containers are small, zeppelin-shaped wicker baskets, of a beauty that is almost surreal. In Toda's notes that describe the materials used in such works, one reads that certain balsa wood surfaces have been inscribed with mesh designs printed by silkscreen using silver-white lead-based oil paint. It is a detail that sounds arcane until one perceives that as it resists the X ray, the lead design articulates the structure of a wooden bar, and similar forms, by giving them a cagelike stereometric projection in the final work.

Toda's practical solution delivers a poetic touch. Such "touches" contribute to the symbolic dualism of the X-ray series in which he contrasts the static, man-made present with a glimpse of organic life that is animate, yet also posthumous in the guise we see.

How often in that X-ray series of thirty-nine works does he employ a flare or fanlike disposition to dramatize an implied motion of the fauna contained in the wooden pieces! His creatures are shown as they would swim, wriggle, or fly—consumed by unearthly white fires that resemble those observed by phototelemetry in the faraway, black heavens. Leafing through the pages of this book, one is invited to reflect on events such as explosions that occurred aeons ago, or on fossilized animal lives in all their graceful, inner filaments, structured by nature even if they only seem dusted onto the black fields. In such manner, Toda layers his X rays with metaphors of remote time.

Occasionally, these metaphors are of recent historical episodes, too, as in the inclusion of the tail assembly of a toy B-29 Superfortress, the plane that bombed Japan in World War II. Toda is not the only Japanese artist to refer in present artistic memory to that charnel moment. The desolation of the war has haunted the career of the senior photographer Shomei Tomatsu, and it figures by implication in the urban ruins photographed by the much younger Ryuji Miyamoto. But perhaps it was left to Toda to juxtapose the fireball of Hiroshima with the theme of the four seasons. In a photography and print project of 1985, titled *First Atomic Bomb*, the seasons are symbolized by frondlike forms made out of transparent resin. For the artist, the resin resembled crystal, which, in effect, is manifested by light. "And the brightest light in the world? The first people to look up and see it were in Hiroshima on that day in 1945."

Earlier, I said that the media constituted a main subject area for American artists, and that their attitude toward the media was nominally critical. Even if they wanted to, how could they avoid dealing with the fact that the everyday American environment was being supplanted by, or rather, was in the act of becoming, a tidal wave of corporate-sponsored imagery? I do not imply by such a statement that the media are any the less of a phenomenon in Japan than they are here. On the contrary, the advance of a postindustrial society in Japan has generated a huge influx of popular imagery, turned out in the usual media: film, TV, comics, magazines, photo-advertising illustrations, and posters. The strength of image culture in Japan has a native origin, even as it also parallels a similar orientation in the United States. Toda's X rays apparently do not pertain to such a vernacular state of affairs, but this may be because he is a worker in the *media* himself, and a considerable one, a longtime prize-winning graphic designer. It would appear that precisely because he knows the commercial and the artistic disciplines at firsthand, and has a firm identity in both, he has kept his practices of them separate from each other.

This is probably true in the technical and stylistic sense, but not always in the iconographic one. Nor in more subtle matters of psychology. My reason for mentioning some of his commercial imagery is once again to profile the X rays, but from a new area, in which craftsmanship is given the highest priority. In his posters for Vivre 21, Isetan, Seio, and Linea Fresca, photography combined with computer-graphic techniques visualize textures of biomorphic shapes that go far beyond the glossy into the slick and fetishistic. Tomato red breastlike forms seem to be made out of the shiniest acrylic; skeletal hands are composed not of bone but of chrome, and they claw the air. It is not merely that natural materials have been replaced by synthetic ones; the mutation occurs in an atmosphere of violence. And the sense of touch is effectively repelled by the cold metallic surfaces of eroticized objects.

I don't know if anyone has commented on the memories of historical trauma that may suffuse these icons that perform a consumer function. Toda manages to suggest uncanny emblems of affluent lifestyle embodied by prosthetic devices! And it is true that his use of X rays for artistic purposes also carries with it an association of symptoms and malaise, if only because we locate the X ray in the medical laboratory. In this context, don't we recall that with their radiation, X rays are actually dangerous to life forms if carelessly handled? Further, his chosen medium in the *X=t* series offers a parallel to that greater radiation released by the bomb itself.

Oddly enough, the closer we look at the *X=t* pictures the more we see how they technically differ from those faint ones produced in the hospital. These, printed here, look in contrast as if they are on pictorial steroids. Instead of the milky dullness, soft contours, and indifferent grays of those wan negatives seen in the consulting room, they are astonishingly crisp, the work of an artist who has placed special demands on the medium. To be sure, whichever his field, the tautness of pictorial control is both highly developed and expensive in its operations. But in the X rays, as one now appreciates, the distancing mechanism of the light-dark reversal is a reflex of a poetic intimacy. It is not a question of displacing materials, but rather of arousing desire for a tactility that is no longer there. The scary illusionism of the commercial work makes me particularly aware of the hieratic format of the X rays. I come back to them thinking of the preference they show for preindustrial aristocratic Japanese culture, which expressed itself in craftsmanly forms of high art.

Without undue insistence, this tone is established throughout Seiju Toda's X-ray project. One of his shows, of 1983, was a reverie on the Heian period (c. 800—1100), a feudal epoch whose court art was frequently concerned with the "passing of earthly beauty." He installed various objects made out of hinoki, a Japanese cypress wood which reacts physically to any change in climate and light. Some of the wooden constructions he has used for his X-ray series are made of hinoki, among other materials. We look back, not upon the little sculptures made of such wood, nor even the secrets they contain, but upon the cryptic white tracery of them, written in the obliviousness of unlighted zones. As it seems, the writing is taken up through the veins and capillaries of insect wings. But I am only superimposing an analogy upon delicate images which their creator—with a droll literalism—has described as never having been seen by anyone before.

$100 = x$

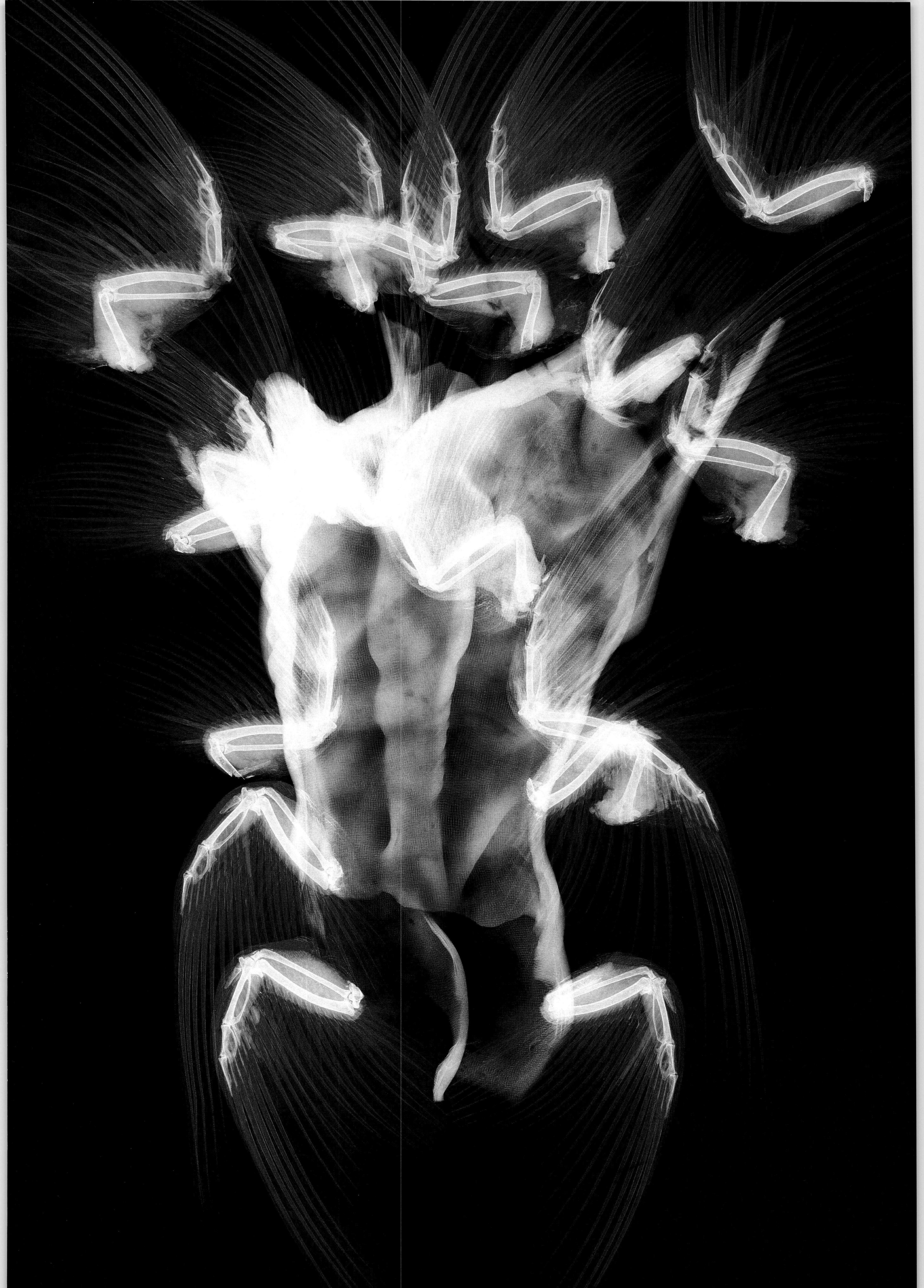

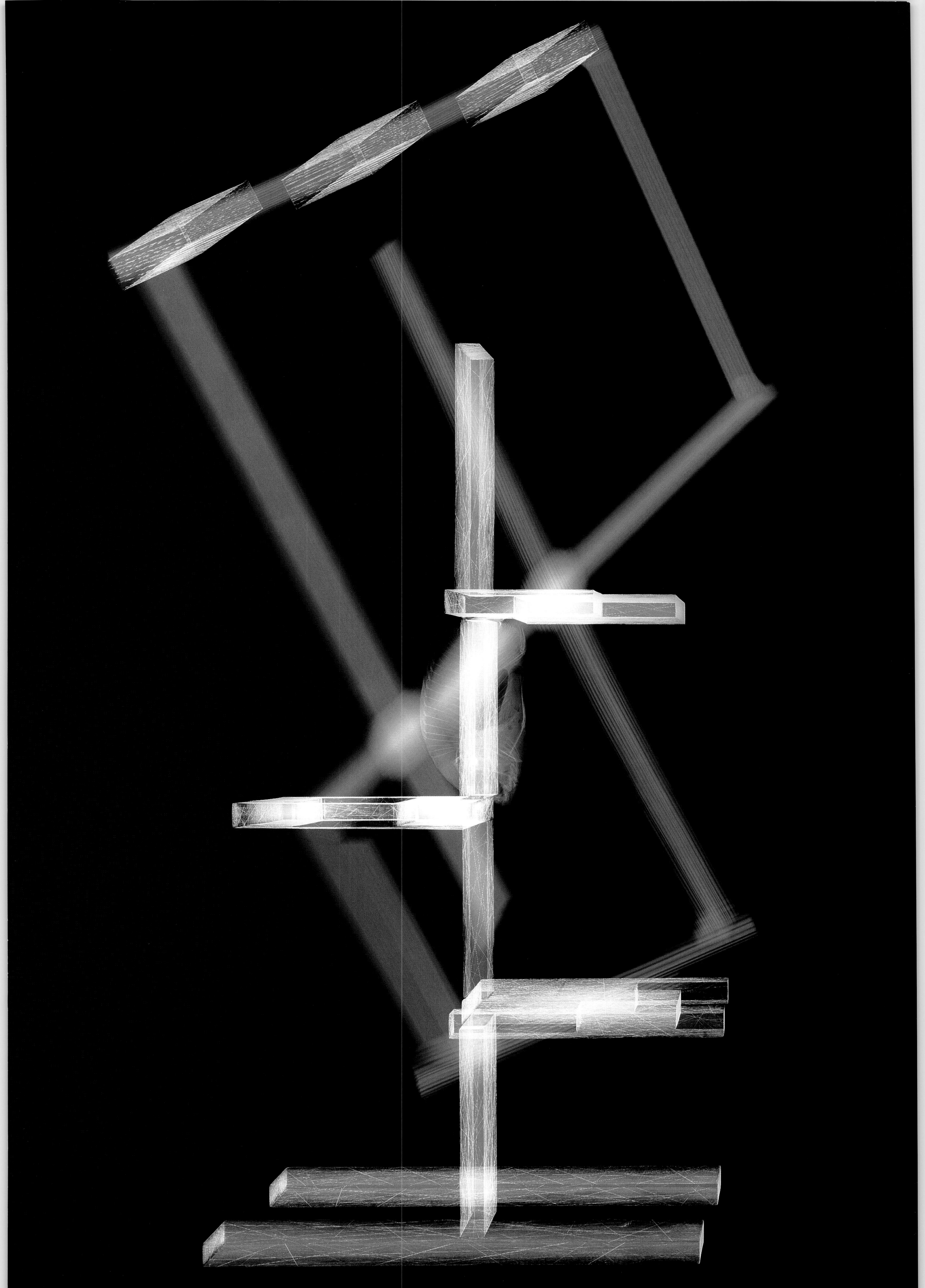

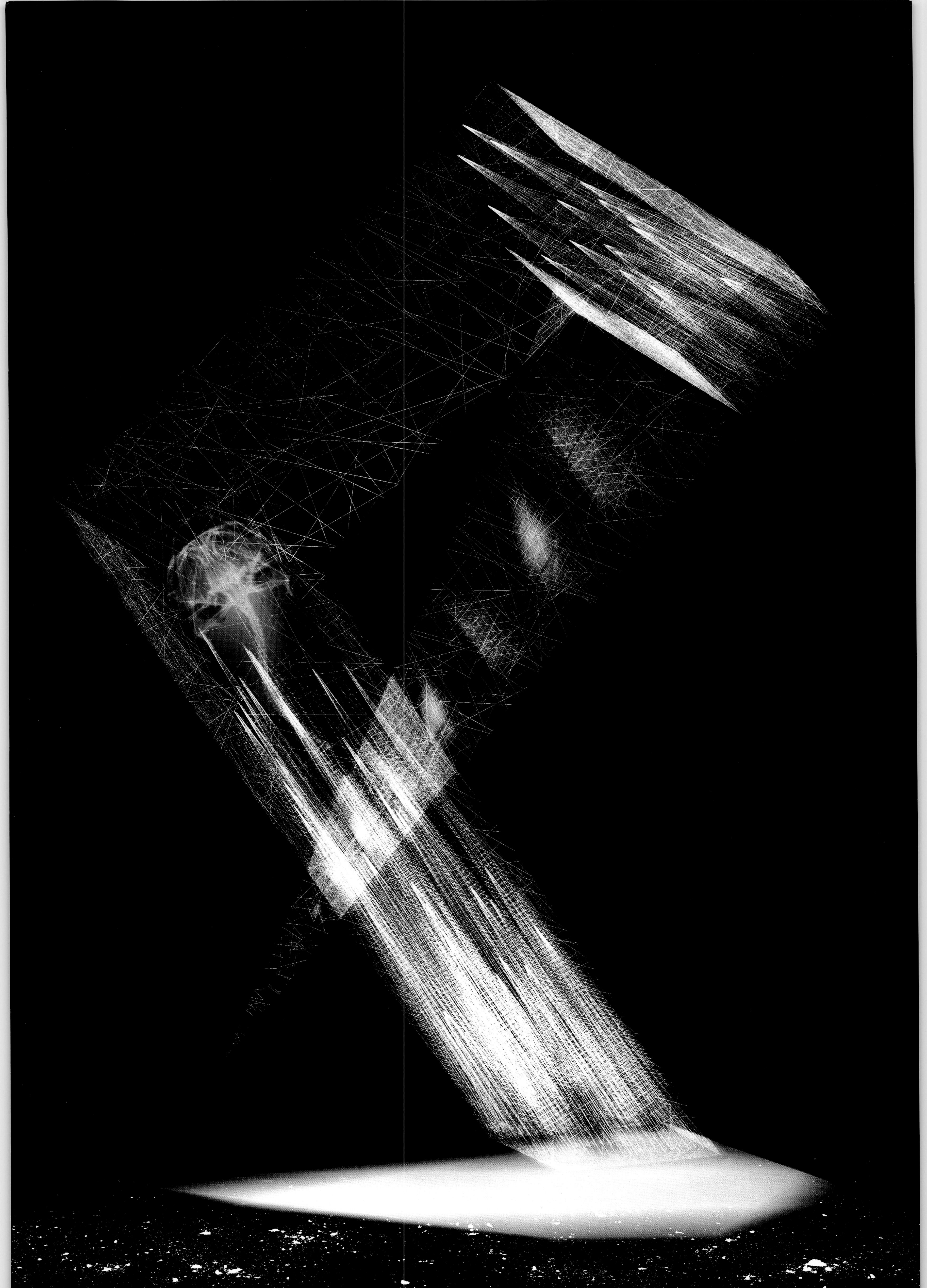

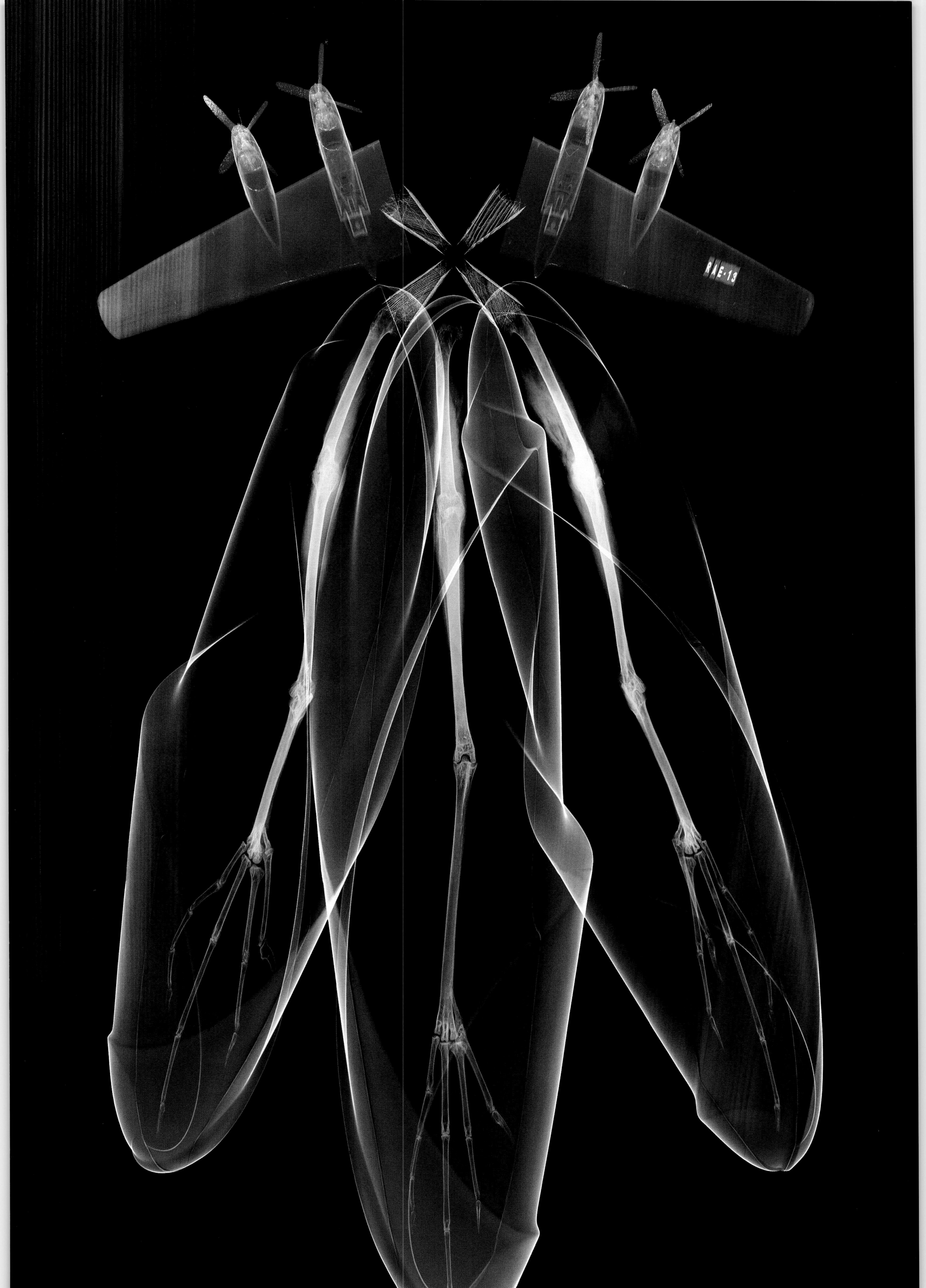
RAE·13

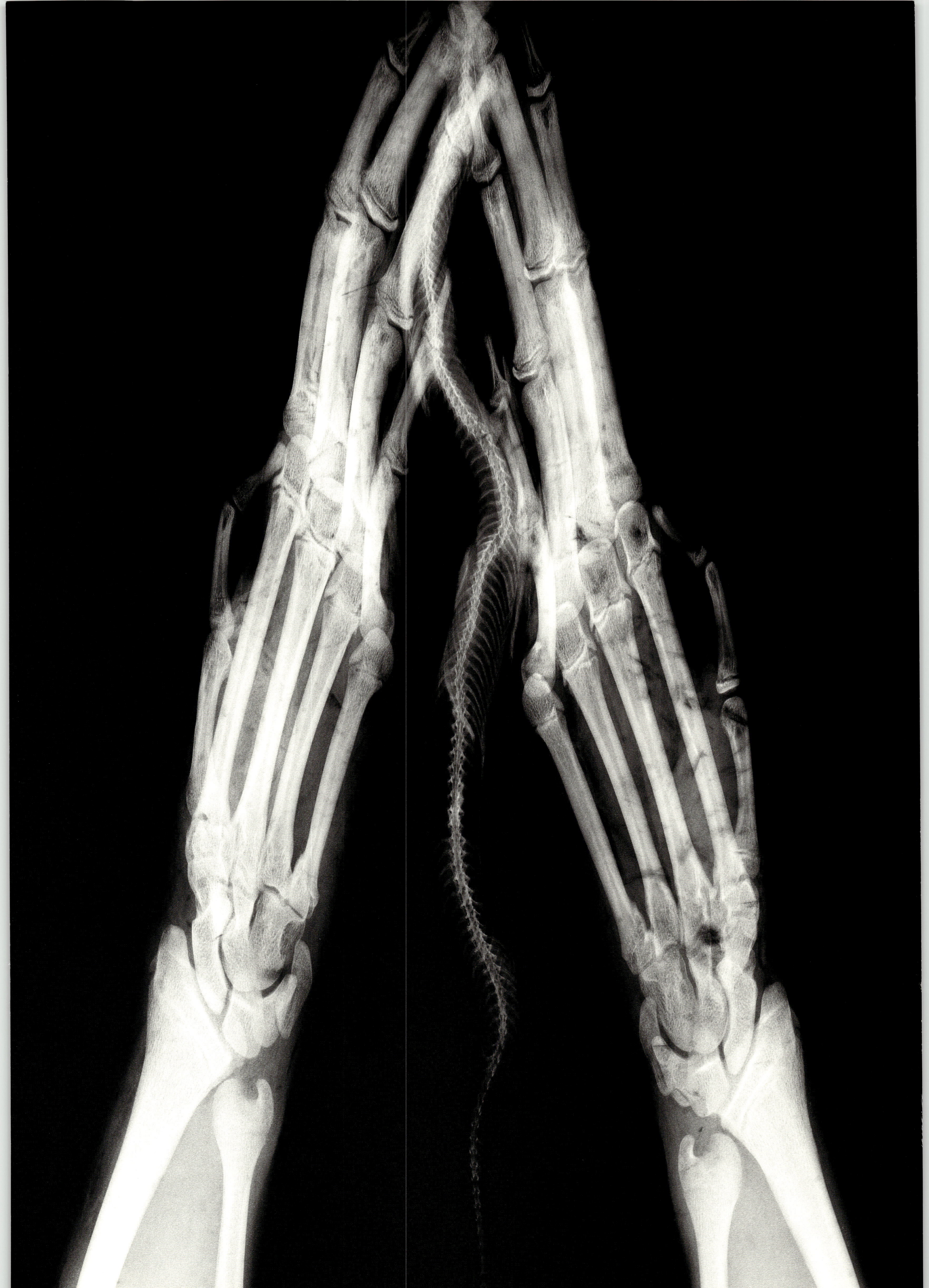

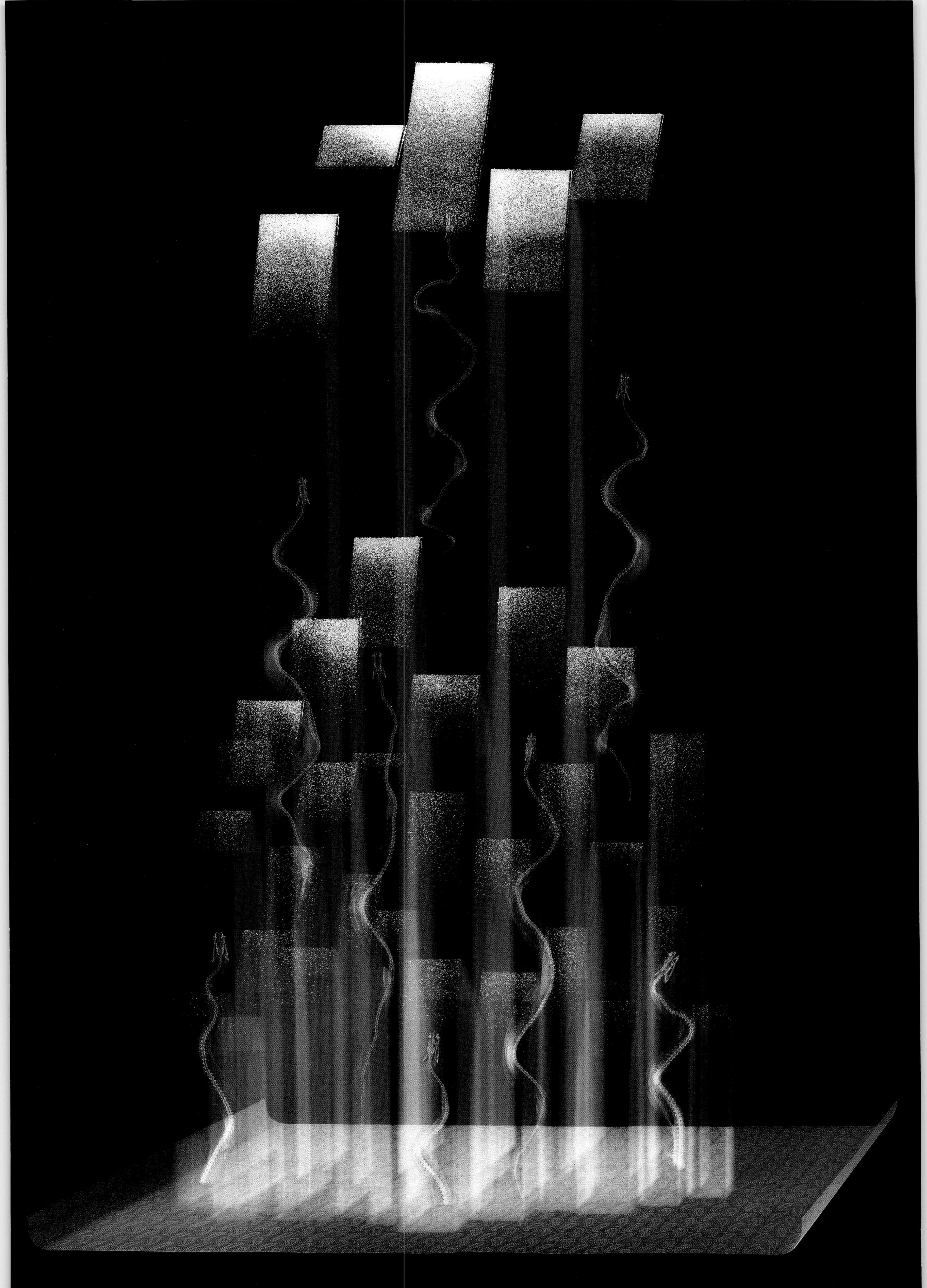

x=019

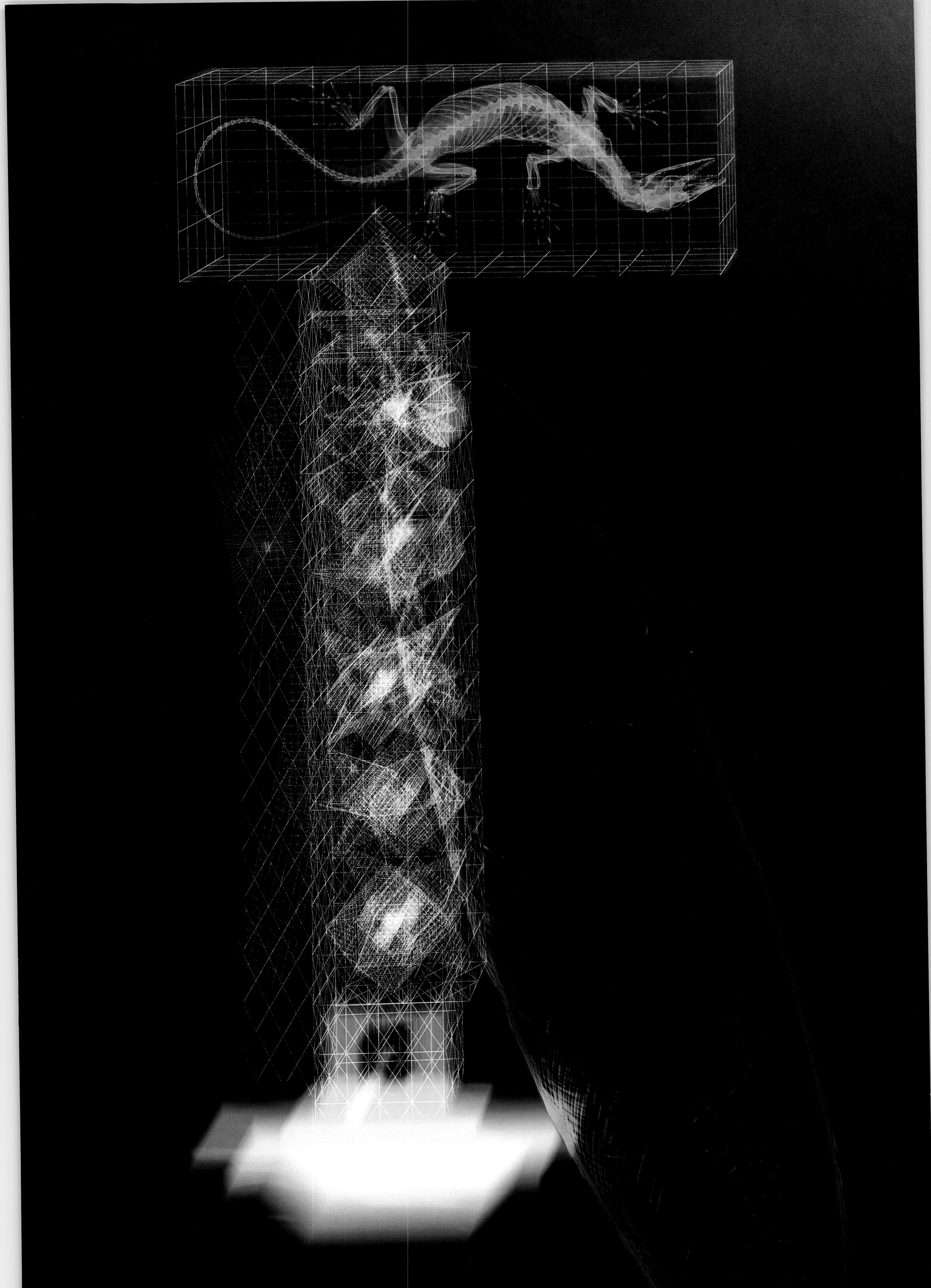

$x=021$

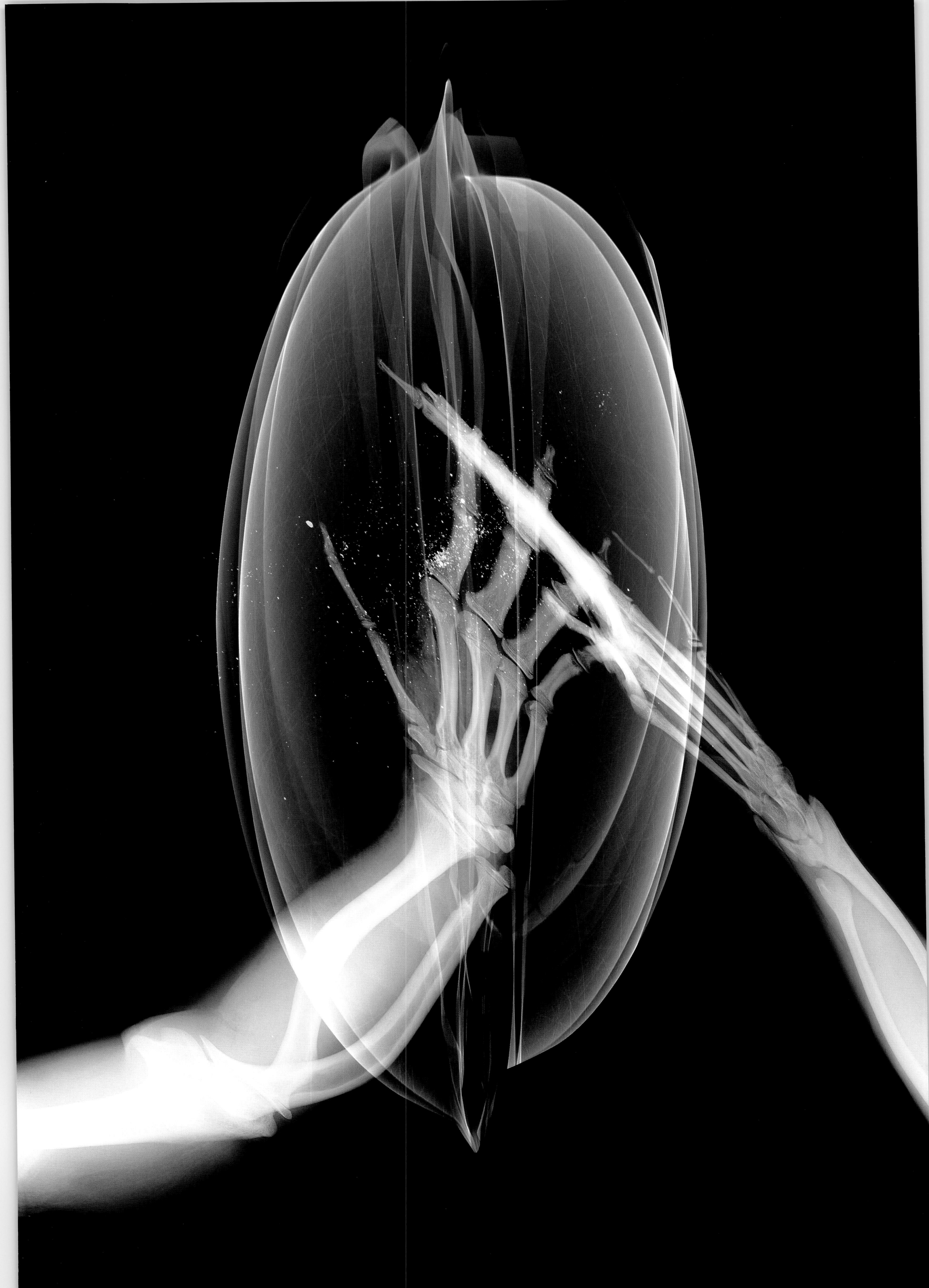

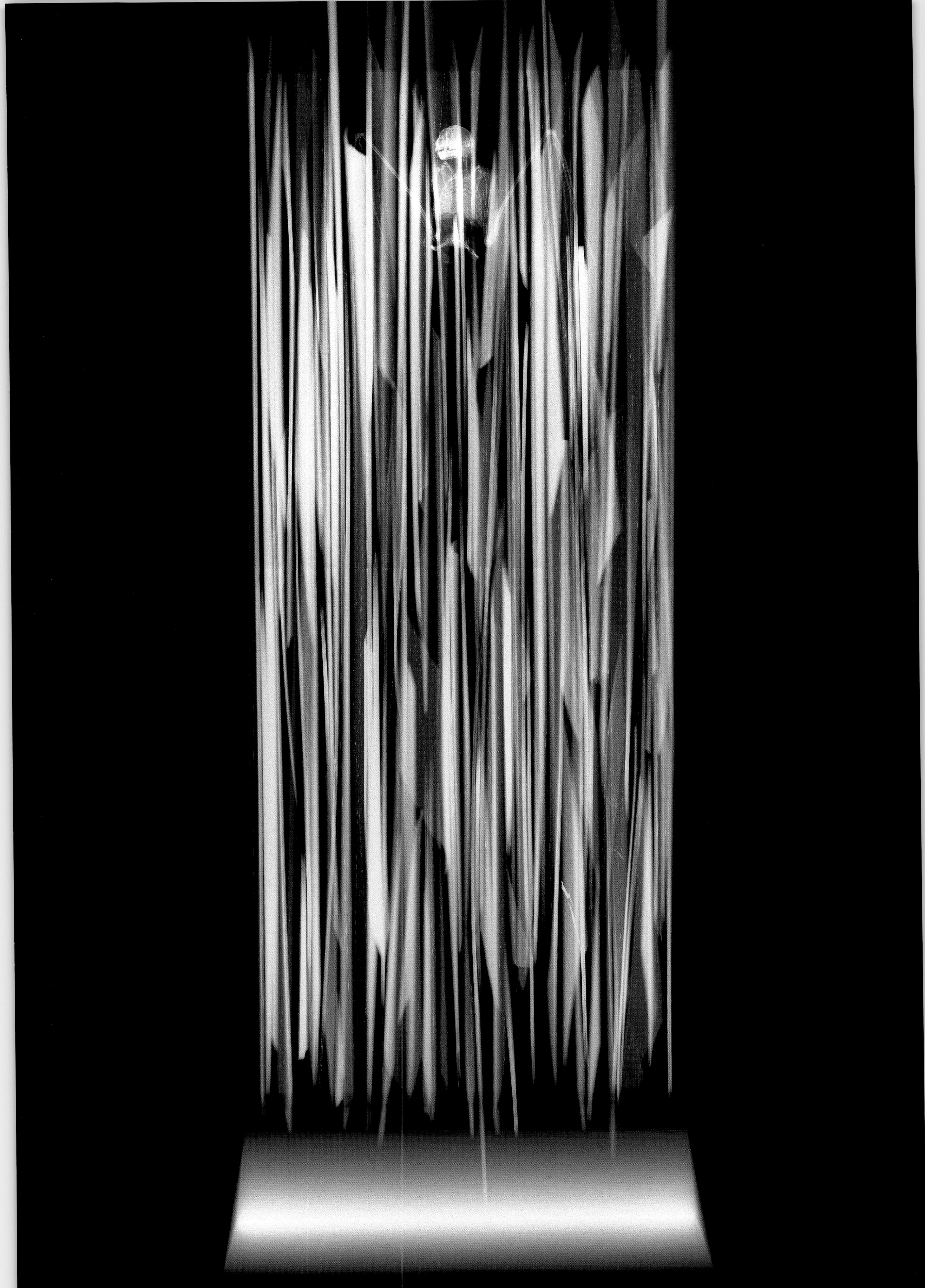

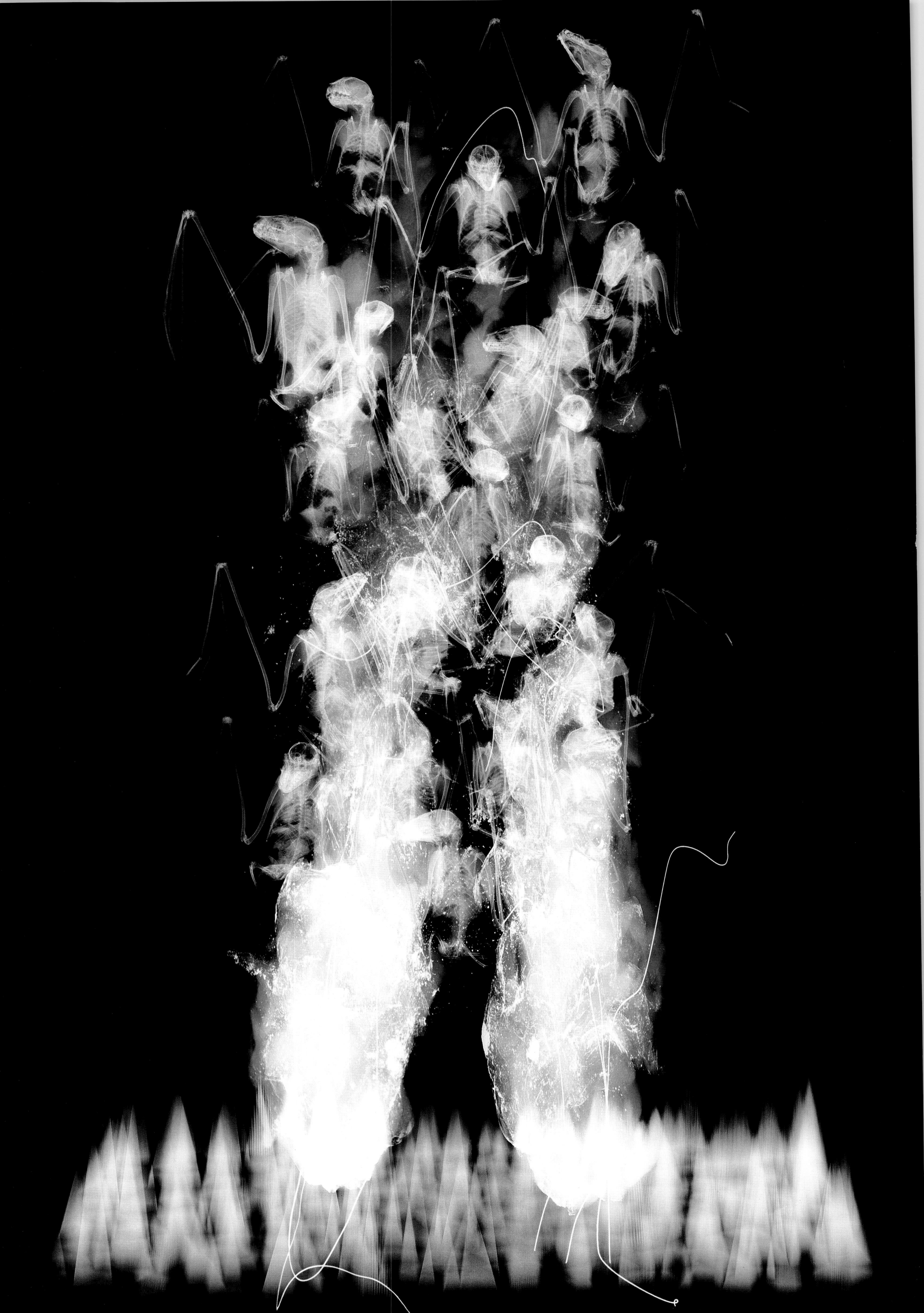

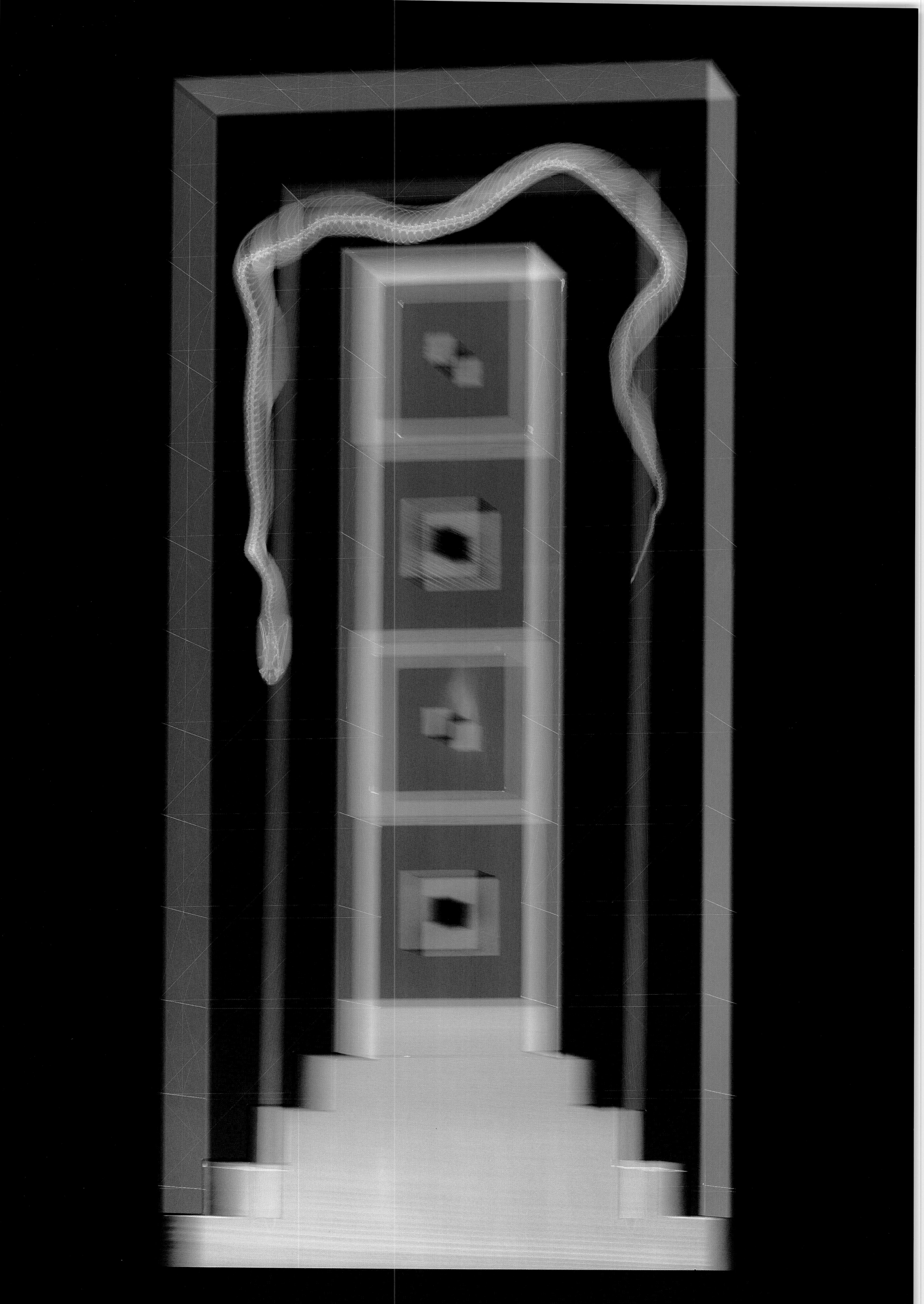

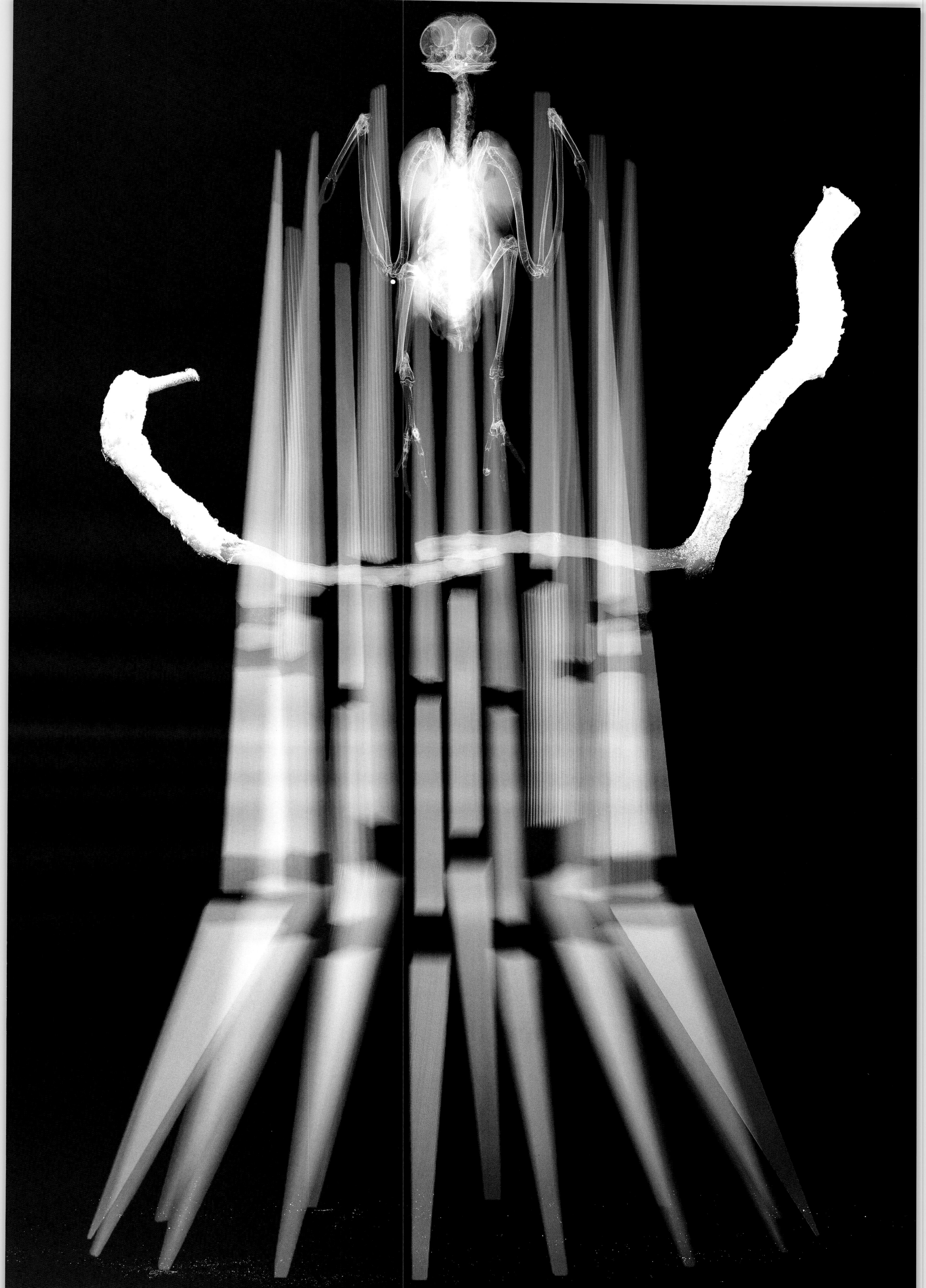

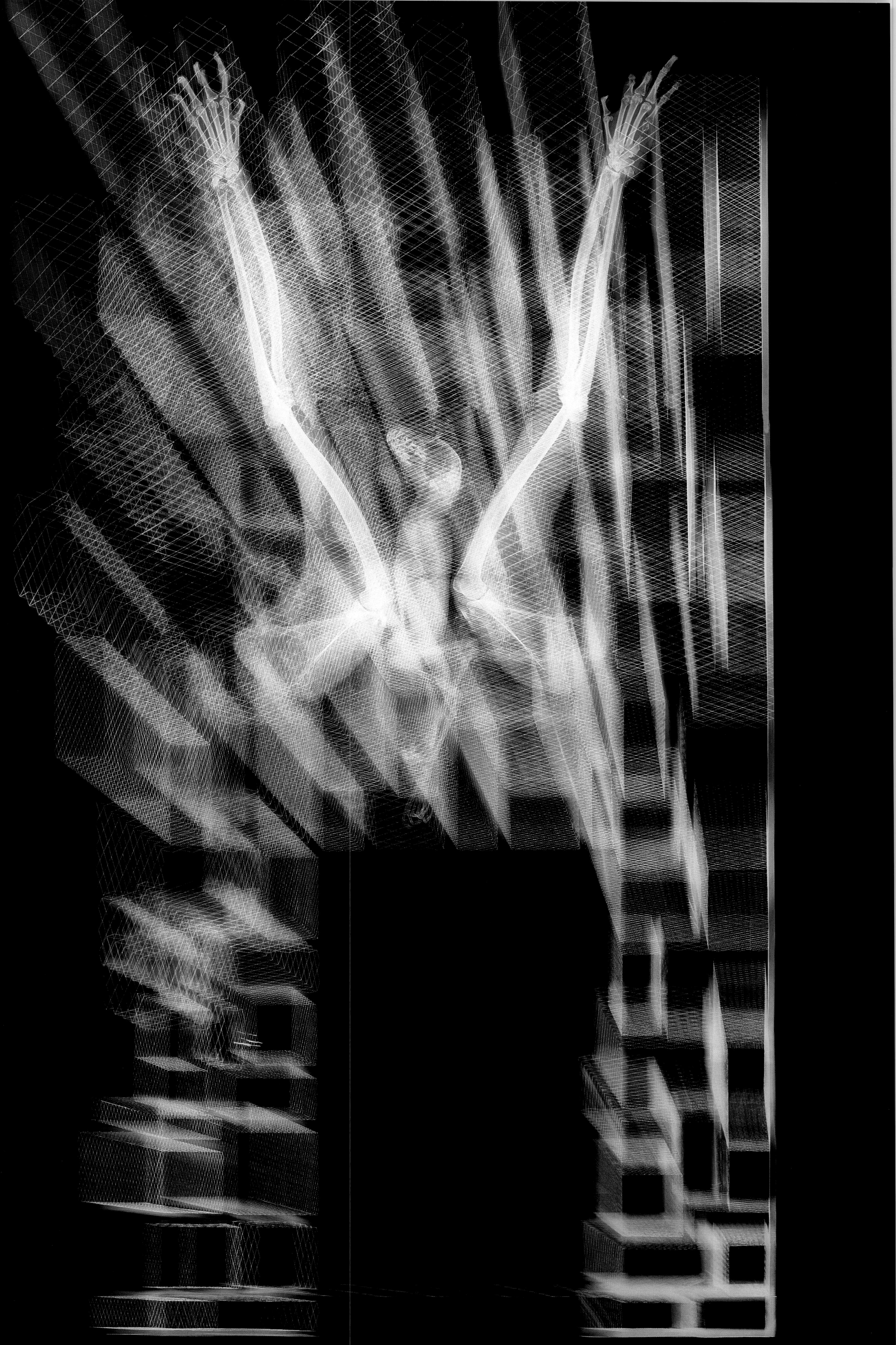

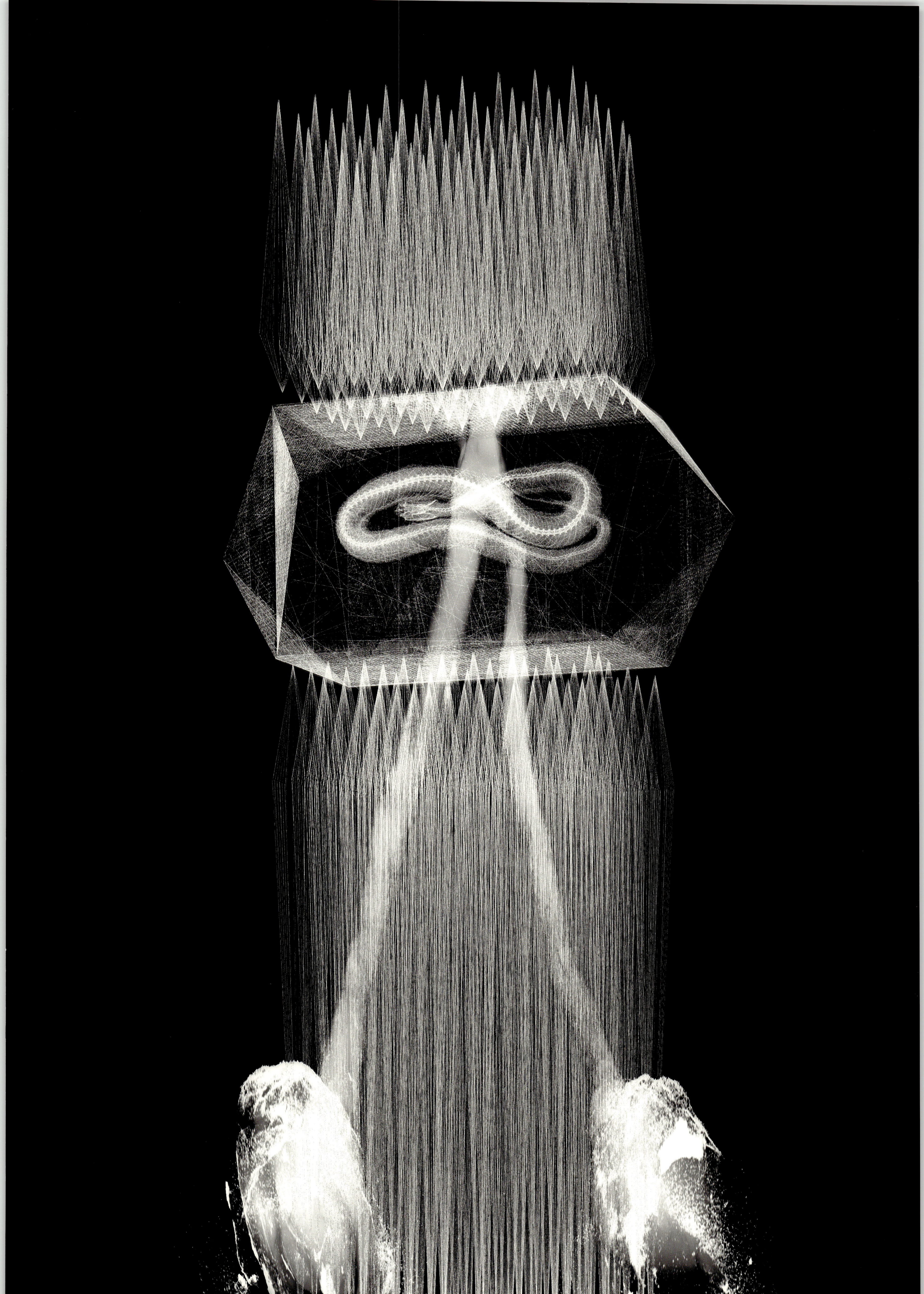

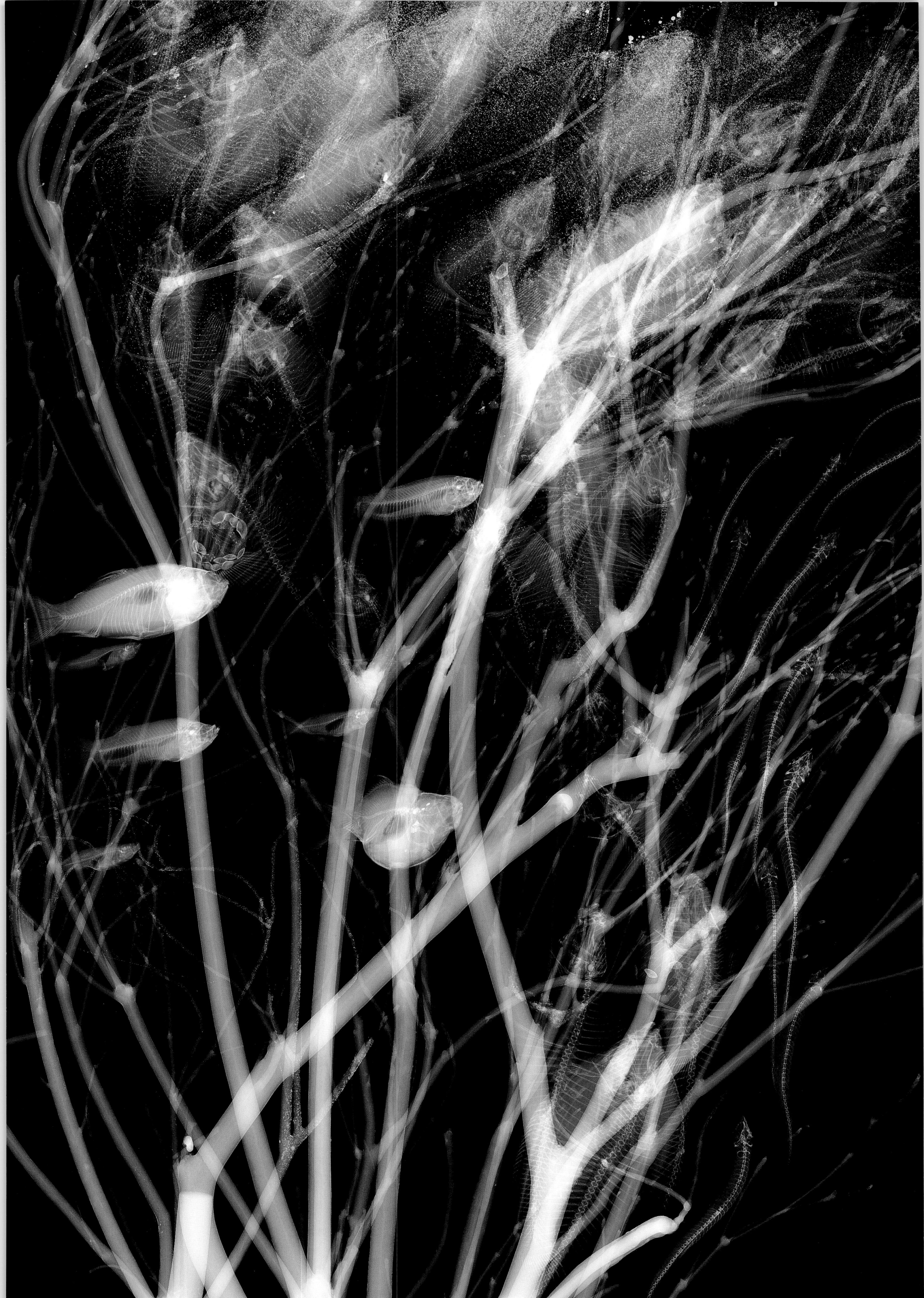

$x=038$

The Installation

At the fountainhead of X-ray photography lies the act of looking at the film itself. A light is transmitted through the film, as in a hospital, where the doctor fixes his gaze at the illuminated film. This is what makes it different from regular photography, where it is the printing that produces the finished form. And this is why this installation was constructed—to enable a reenactment of the first moments in the production of X-ray art. The films are enlarged or reduced for the desired effect, then formed into a collage upon a huge light box on the wall. Then it is the flood of light that makes it all happen.

The installation emits light into the darkness of night. It breathes new life into the murkiness of what appear to be abandoned, dusty, and dirty warehouses.... And we see that it is this light that gives the X-ray photograph its very life.

Photo by TAKAO IKEJIRI

Description of the Plates

Two major premises from the basis for the photo series, *X=t*.

One of these is the idea itself—the truly innovative approach to art involving the design of architecture, the construction of miniature models, and the use of an X-ray camera to capture, as if by looking into the inner souls, what is normally far beyond the visual range of the human eye.

The other is the rendering of an overall image, or a special ambiance, to this photo series. A storyline is given to the program specifically to create the impression of walking through an actual setting and looking at each structure as one goes along.

Drawing the architectural plans entailed considerable time and effort. The earlier works, in particular, were drawn on drafting boards carried into various hotel rooms on my travels—in Sydney, Hawaii, or Paris—far removed from the daily routine of life.

The story evolved out of a human motif of a man and a woman, and expanded to include the sights of a city, typified by the squares and the high-rise buildings, and proceeded to encompass such environments as under the sea and in the air, passing through graves and reaching the realm of outer space for a promise of the future in the final chapter.

What follows in the ensuing pages are the X-ray photographs, together with my own texts on the concepts behind each work.

x=001 The human being as the starting point. Though the outward appearance suggests the enticing body of a voluptuous woman, the X rays reveal the insides to be those of a fractured angel. Her wings lay scattered nearby.
Nude torso of a woman in polyester (with fiberglass); 19 pigeon wings.

x=002 The human being—a man—as the starting point. Signifying destruction of an exuberance for life, the exterior sports a powerful masculine body. This work depicts motion, rather than serenity. Nude torso of a man in polyester resin (with fiberglass); paper clay; 18 pigeon wings.

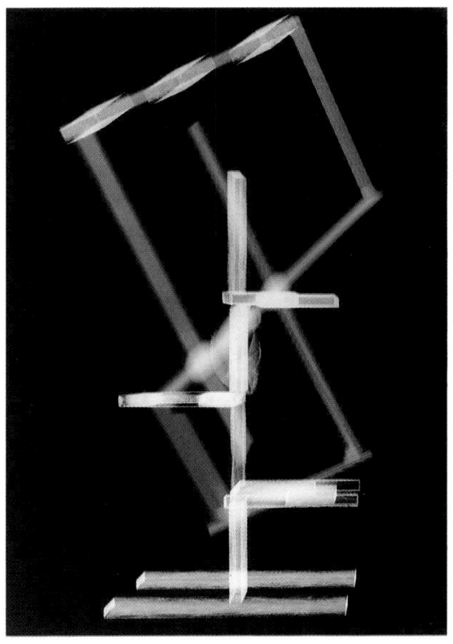

x=003 The first piece of architecture that catches your eye as you enter the town square is the monument. The monument here is made of wood. Under the wooden surface is a fluid movement where a sea bream is trying to swim.
Spruce (wood pieces are randomly scratched with the blade of a retractable knife. Silver-white oil paint containing lead is rubbed into the scratch marks); white ash (rubbed in three places forming bands, with silver-white oil paint containing lead); 1 *ibodai* (a kind of sea bream).

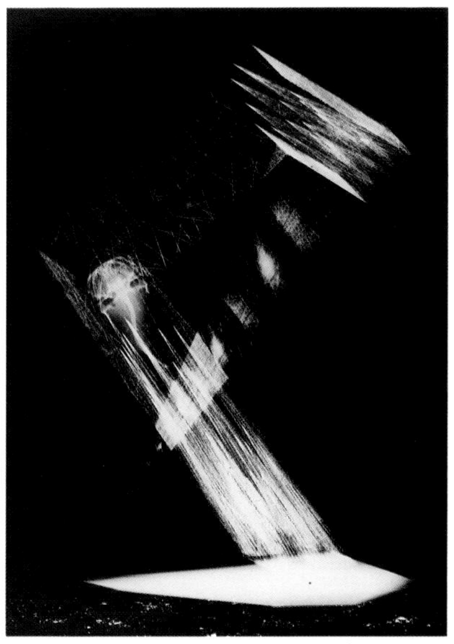

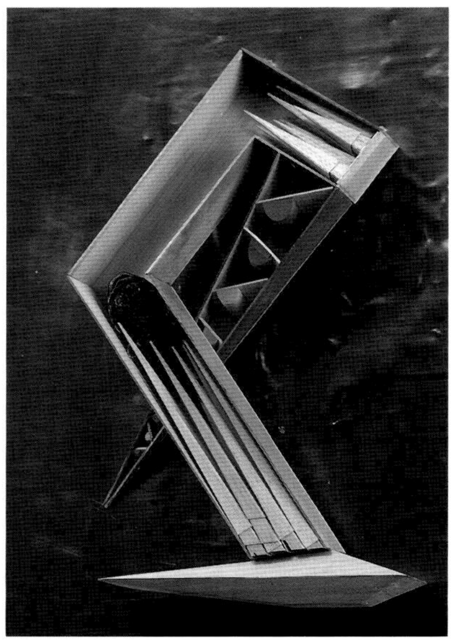

x=004 The idea of fish swimming inside architectural structures is not confined to something as symbolic as a monument. As with all the structures that follow, imagine, if you will, the model expanded to its full scale, and constructed on a solid piece of ground. Let the images reflected in the viewers' eyes assume grander proportions. *Agachisu*, a kind of luan wood (blade-scratched with leaded silver-white oil paint rubbed into the scratch marks); balsa wood (lines printed by silkscreen using leaded silver-white oil paint mixed with ink); spruce; leaded powder pigment; 1 angler.

x=005 A three-legged robot. The concept of mobile architecture recurs elsewhere in the series. These are not made merely to move around in the mind of the artist, but are presented as feasible images desirable for reproduction in real life. Plastic toy-model fighter airplane wings; leaded powder pigment; wire; lead letterings; spruce (blade-scratched with leaded silver-white oil paint rubbed into the scratch marks); 3 twisted ellipsoids of vinyl chloride; 3 heron legs.

x=006 Not only are there buildings that walk; here is one that may very well fly. Large passenger planes that cover the continents are, in effect, "home" for hundreds of people during their flights. *Agachisu* (blade-scratched with leaded silver-white oil paint rubbed into the scratch marks); plastic toy-model fighter airplane (rear segment); spruce (blade-scratched with leaded silver-white oil paint rubbed into the scratch marks); paper clay; 8 mamushi (pit vipers).

x=007 A huge building, supported by two legs. Architecture, in this context, can be likened to sculpture. *Agachisu* (blade-scrached with leaded silver-white oil paint rubbed into the scratch marks); cherry branches; spruce; leaded powder pigment; 2 mamushi.

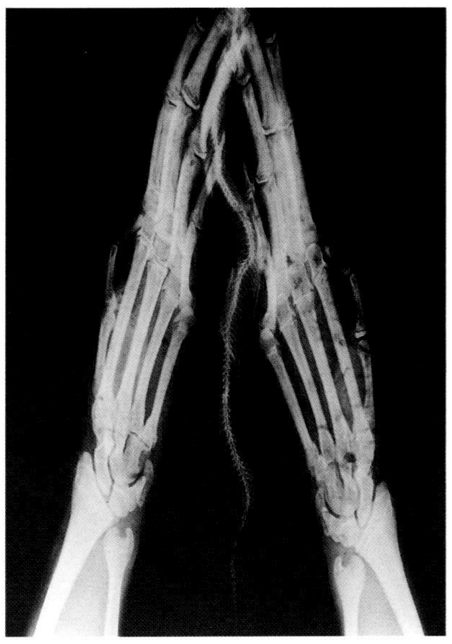

x=008 Praying for peace. Buildings symbolize impermanence for all of them are demolished sooner or later. As such, they invariably house a pair of hands joined in prayer. One pair of chimpanzee hands; 1 mamushi.

x=009 A windmill with hand-sails radiating from the apex of a pyramidal windshaft. Hinoki (Japanese cypress); spruce (blade-scratched with leaded silver-white oil paint rubbed into the scratch marks); leaded powder pigment; 4 chimpanzee feet.

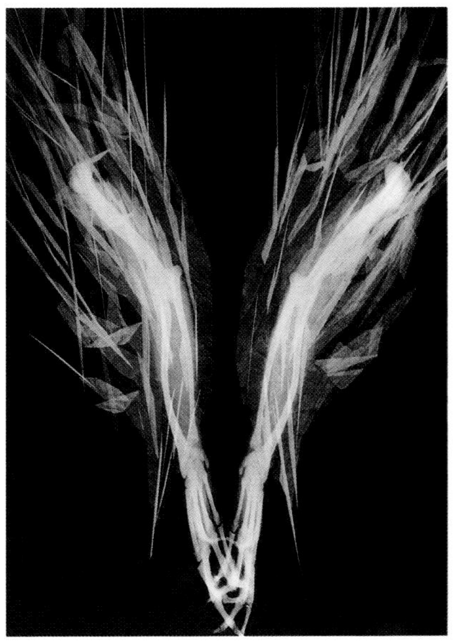

x=010 Arms take the role of wings and clasped hands act as the hinge for a rendition of a building in flight. Plywood (blade-scratched with leaded silver-white oil paint rubbed into the scratch marks); one pair of chimpanzee arms.

x=011 A cluster of towering buildings representing futuristic housing. In reality, each tower contains only one apartment regardless of its height. The apartment buildings offer a choice of how high one wishes his space. Each tower contains one finger. *Agachisu*; wooden *agachisu* disk; leaded powder pigment; 5 chimpanzee fingers.

x=012 The same design of building as in No.11. In this version, the presence of pit vipers converts the concept of height governing interior space to the concept of length. *Agachisu*; leaded powder pigment; vinyl sheets (paisley patterns printed by silkscreen using leaded silver-white oil paint mixed with ink); 8 mamushi.

x=013 Fastened above a gigantic building, an even larger mobile of baskets hovers in the wind. Inside, the ugly-looking stingfish swim toward the skies. All the creatures in this photo series tend to look hideous on the surface, but how lovely are their bones ! 6 baskets; balsa wood (lines printed by silkscreen using leaded silver-white oil paint mixed with ink); hinoki cypress; spruce; paper clay; acrylic rods; 3 stingfish.

x=014 A building that blossoms like a flower. 9 baskets; hinoki cypress; acrylic rods; paper clay; balsa wood (line printed by silkscreen using leaded silver-white oil paint mixed with ink); spruce; 8 stingfish.

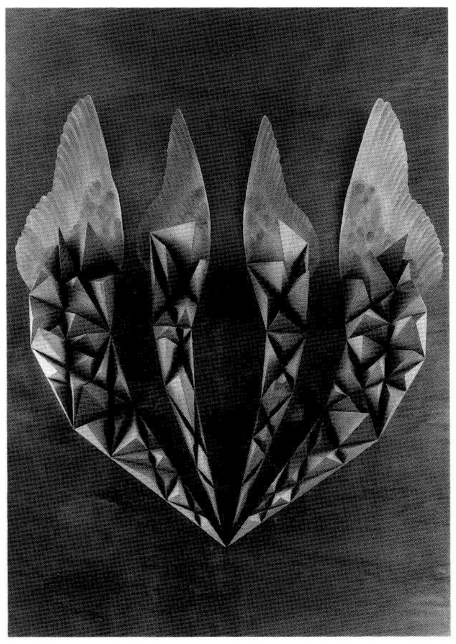

x=015 Praying with four hands. The X-ray camera transforms the image into a dazzling chandelier. The chandelier resembles the UFO of "Close Encounters . . ." The overpowering brilliance also reminds mankind of the atomic bomb dropped on Hiroshima. In the blinding light, a church—or a shrine—prays for peace. *Agachisu*; spruce in triangular and quadrangular pyramidal pieces; 4 pigeon wings.

x=016 An image in the same sequence as No. 15. A prayer derives its significance from continuation. *Agachisu*; spruce in triangular and quadrangular pyramidal pieces; 3 pigeon wings.

x=017 A building constructed upon the ocean floor. A veritable paradise with tropical fish, this is a world with the kind of peace and tranquility that we have yet to establish in our world. Spruce; cherry branches; 5 fighting fish (tropical fish).

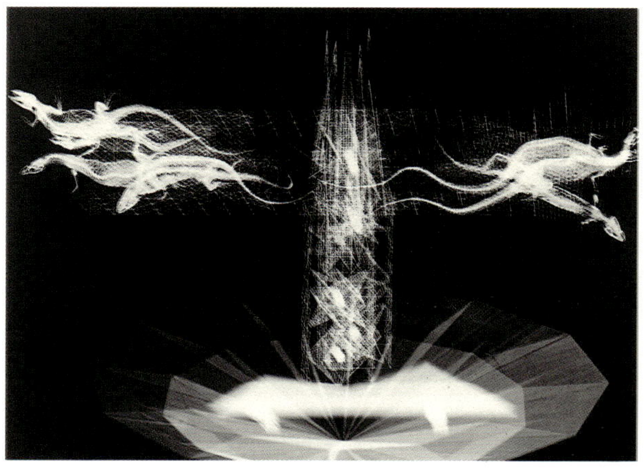

x=018　A home for lizards. The reptiles boast the longest history of existence on Earth. What would show up on their DNA printouts?　Balsa wood (lines printed by silkscreen using leaded silver-white oil paint mixed with ink); spruce; wooden *agachisu* disk; 7 Timor monitors (lizards).

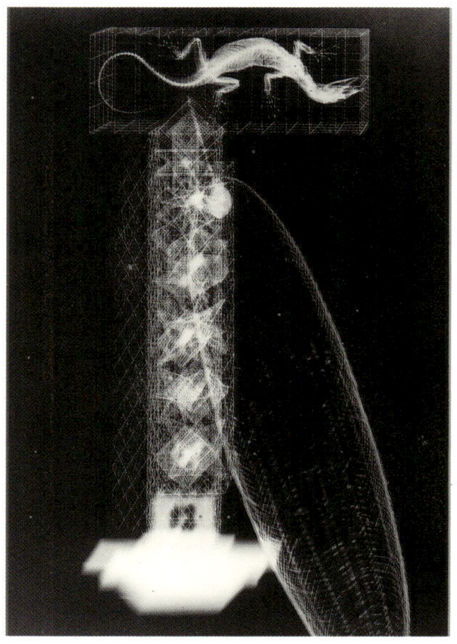

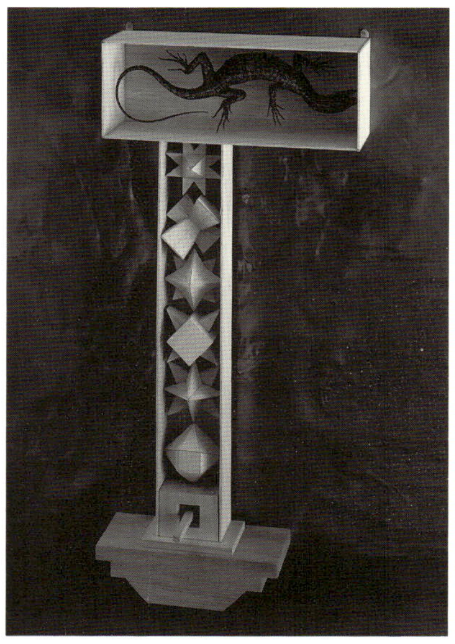

x=019　The image of a house of death. Here is a grave of a lizard reaching out toward heaven.　Balsa wood (lines printed by silkscreen using leaded silver-white oil paint mixed with ink); spruce; 1 basket; hinoki cypress; 7 Timor monitors (lizards).

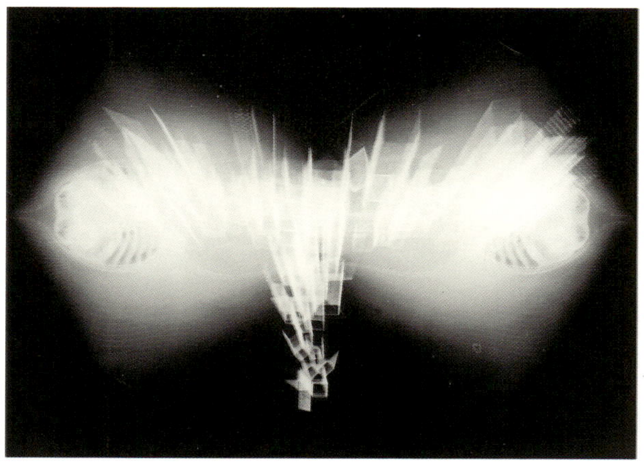

x=020　The bones found in nature compose the loveliest structures in the world. Hard as they are, they are still capable of lissome movements, as shown here in an image of a delicate machine.　Balsa wood (lines printed by silkscreen using leaded silver-white oil paint mixed with ink); 2 stingrays.

x=021　The stingray, which inhabits the ocean depths, is, in a way, a seabird with wings. Here, the seabird takes flight toward the skies.　9 sculpted pieces in vinyl chloride (blade-scratched with leaded silver-white oil paint rubbed into the scratch marks); leaded powder pigment; 1 stingray.

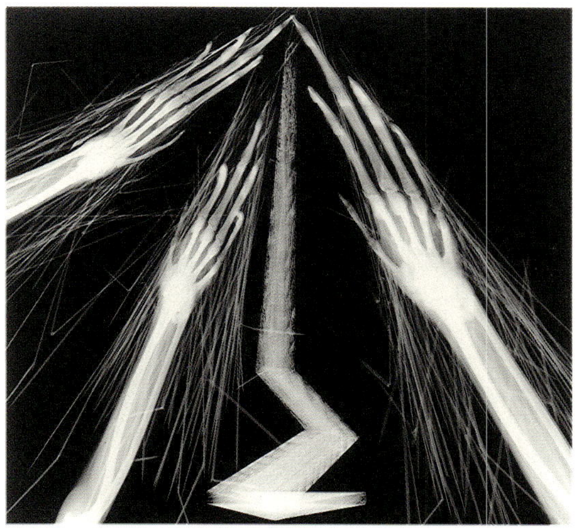

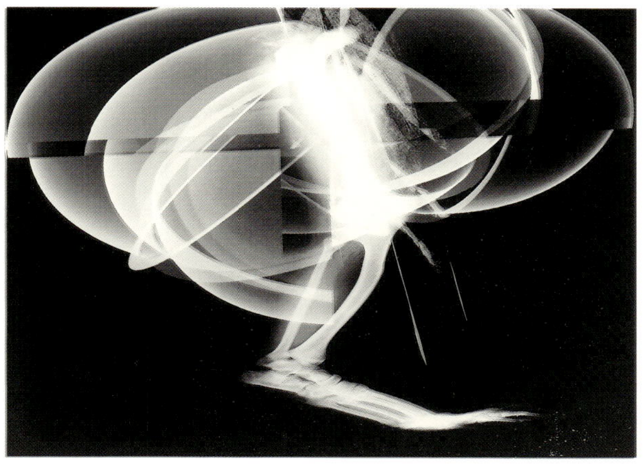

x=022 A structure inspired by Michelangelo's "Creation of Adam," in which the outstretched fingers point toward each other. Here, three hands created by nature and one man-made hand together extend toward a point of convergence. Hinoki cypress; spruce (blade-scratched with leaded silver-white oil paint rubbed into the scratch marks); 3 chimpanzee hands.

x=023 No reason why there should not be a piece of architecture that moves. Rimmed half-spheres in vinyl chloride (blade-scratched with leaded silver-white oil paint rubbed into the scratch marks); plywood (similarly blade-scratched and rubbed with paint); leaded powder pigment; 1 chimpanzee foot.

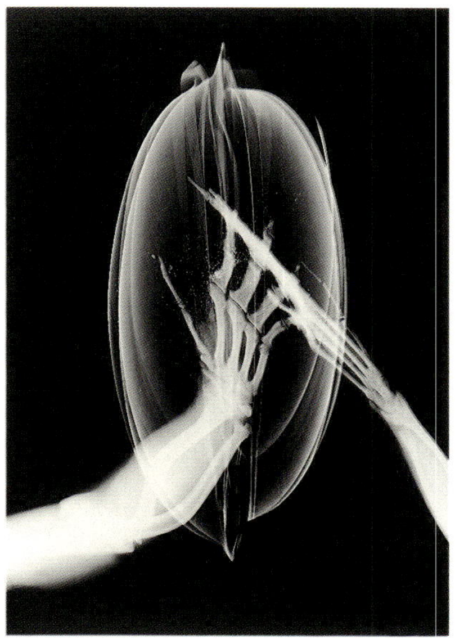

x=024 A man's hand and a woman's hand entwine in love inside an artificial uterus. Here, the architecture is a symbol of sex. Rimmed half-spheres in vinyl chloride (blade-scratched with leaded silver-white oil paint rubbed into the scratch marks); leaded powder pigment; 1 pair of chimpanzee arms.

x=025 An ugly—and beautiful—moray eel forms the figure X. The image is that of the earth before anything is built, where we find rot, wasteland, the wind, and all kinds of rubble. Paper clay; plywood (blade-scratched with leaded silver-white oil rubbed into the scratch markes); leaded powder pigment; wire; 1 moray eel.

x=026 A grave for a feather. The enveloping silence reigns inside the grave as well. *Agachisu* (blade-scratched with leaded silver-white oil paint rubbed into the scratch marks); white ash (rubbed in three places forming bands, with leaded silver-white oil paint); spruce (similarly scratched and rubbed with oil paint); wooden *agachisu* disk; 1 pigeon wing.

x=027 The corpse of a ship. It sleeps in its final resting place under the sea. Paper clay; plywood (blade-scratched with leaded silver-white oil paint rubbed into the scratch marks); plastic toy-model battleship; leaded powder pigment; wire; 1 moray eel.

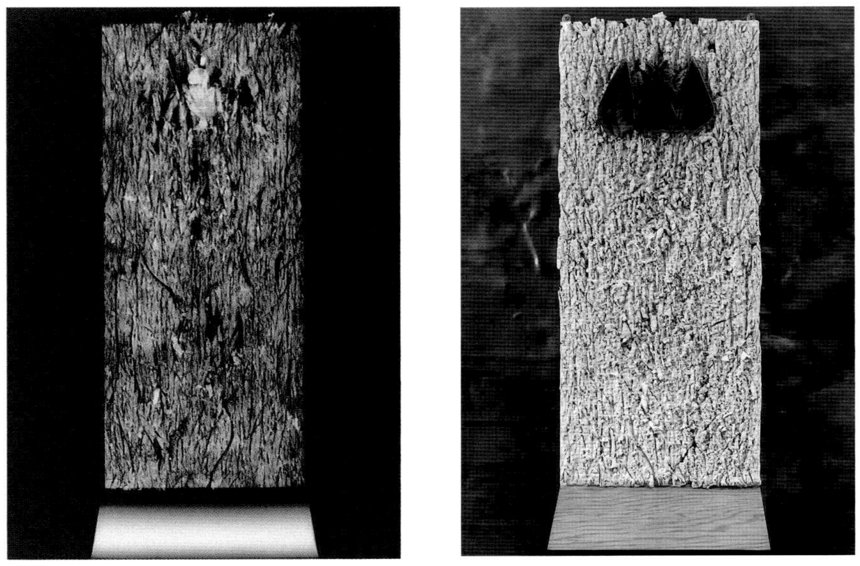

x=028 The home of a bat. An abode, in life, for a creature exuding vitality. Plywood (blade-scratched with leaded silver-white oil paint rubbed into the scratch marks); paper clay; spruce; 1 bat.

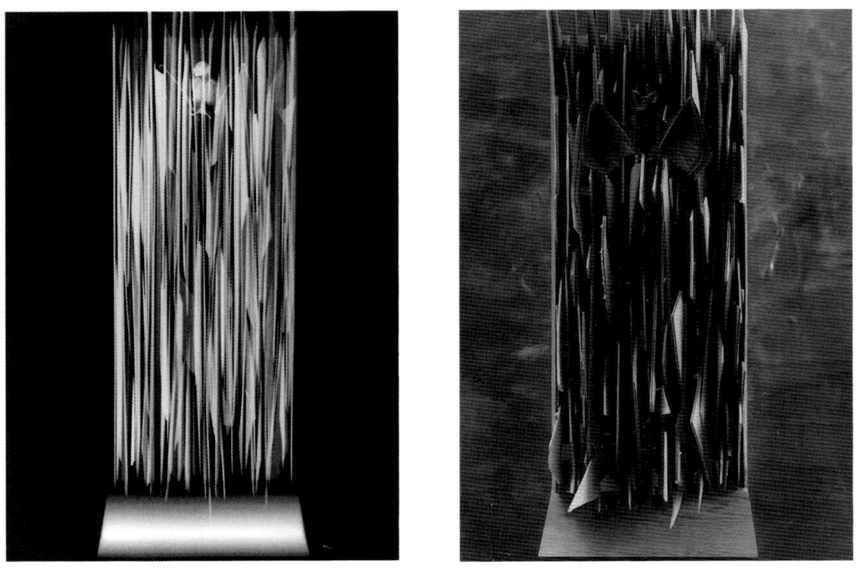

x=029 The crucifixion of a bat. The cross provides an abode, in death. A bat is a mammal that sees dreams in the darkness, thereby resembling the human being. Plywood (blade-scratched with leaded silver-white oil paint rubbed into the scratch marks); spruce; 1 bat.

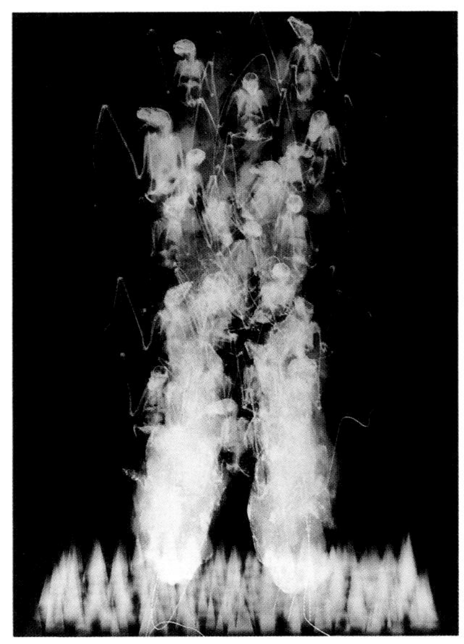

x=030 In the sequence of things, after life and death comes birth. This is the bat's abode, in birth.
Paper clay; leaded powder pigment; wire; spruce in quadrangular pyramidal pieces; 21 bats.

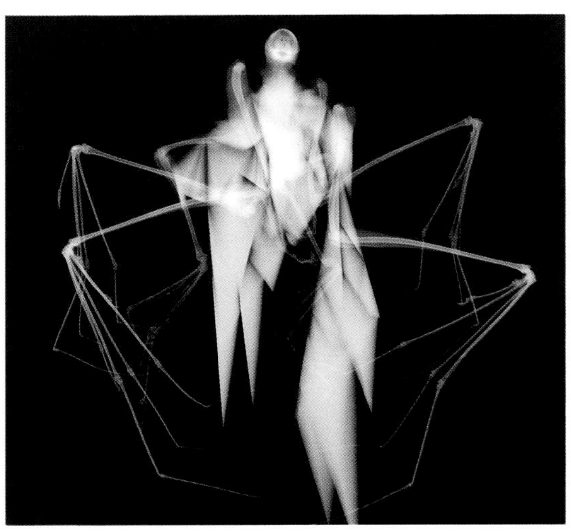

x=031 A composite of a bat and bat wings in a self-portrait of the artist. The imagery consists of four hands deep in thought. One real pair of hands plus two hands of the "alter ego." *Agachisu*; spruce in triangular and quadrangular pyramidal pieces; 1 bat; 1 pair of bat wings.

x=032 The disembodied spirits of bats; in a wavering, meandering structure. 6 baskets; acrylic rods; spruce; balsa wood (lines printed by silkscreen using leaded silver-white oil paint mixed with ink); paper clay; 4 bats without wings.

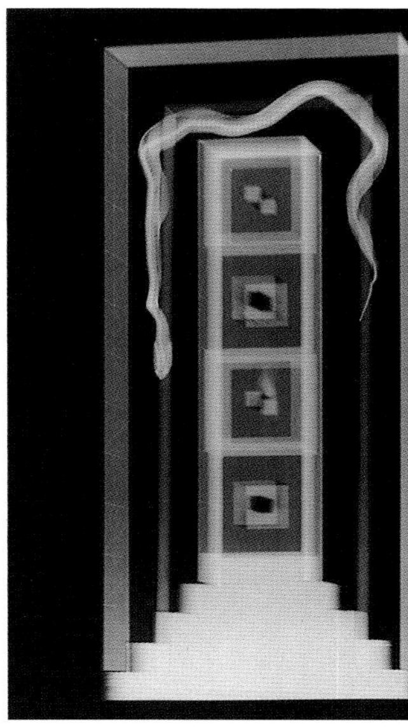

x=033 A building with mamushi (pit vipers), symbolizing the city's paradise lost. *Agachisu* (blade-scratched with leaded silver-white oil paint rubbed into the scratch marks); spruce; 1 mamushi.

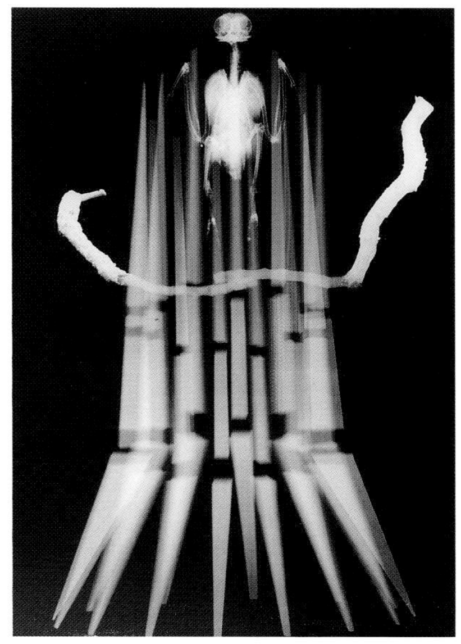

x=034 An owl's home in a tree, symbolizing the city's forest dwellings. The owl watches over human beings through deific eyes. Hinoki cypress; cherry branches; leaded powder pigment; 1 owl.

x=035 A bat and a monkey fuse in a maniacal piece of architecture, symbolizing the city's dwelling for the demons. Balsa wood (lines printed by silkscreen using leaded silver-white oil paint mixed with ink); 1 bat without wings; 1 pair of monkey arms.

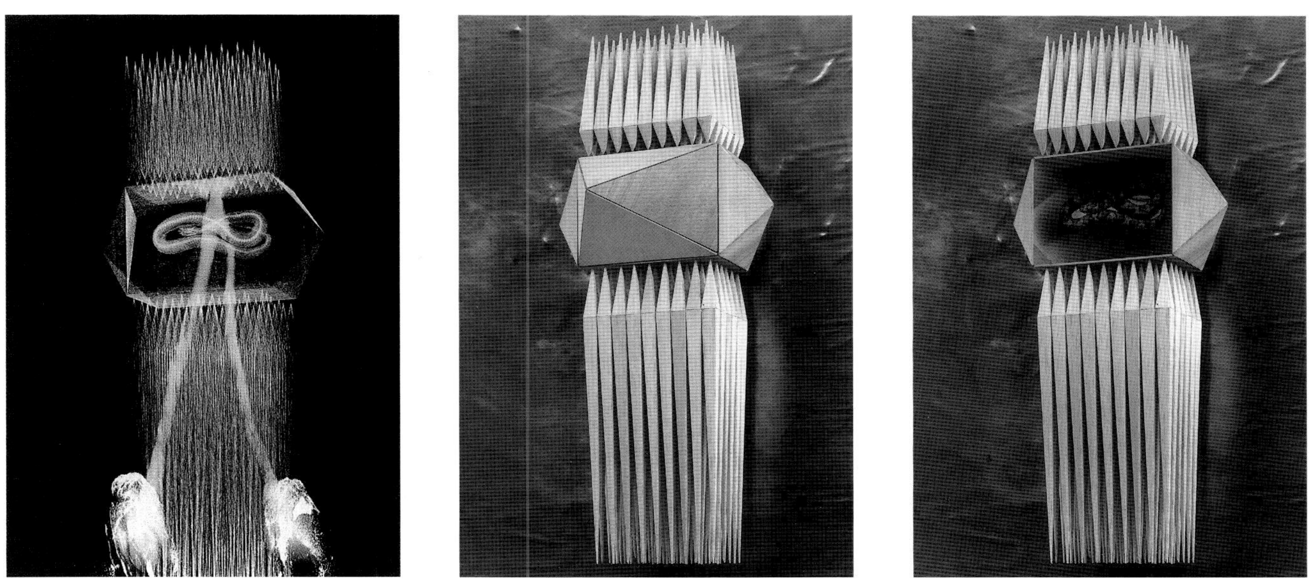

x=036 A building with legs. Though it is equipped with many legs, there is no means of exit. In other words, anyone living inside cannot step out to go anywhere. Here is constraint under the name of freedom. *Agachisu* (blade-scratched with leaded silver-white oil paint rubbed into the scratch markes); balsa wood (lines printed by silkscreen using leaded silver-white oil paint mixed with ink); paper clay; leaded powder pigment; 1 mamushi.

x=037 A forest of trees grows deep under the sea in stormy waters, while tranquility reigns on the surface. Witch hazel branches; leaded powder pigment; flounders; loaches; goldfish.

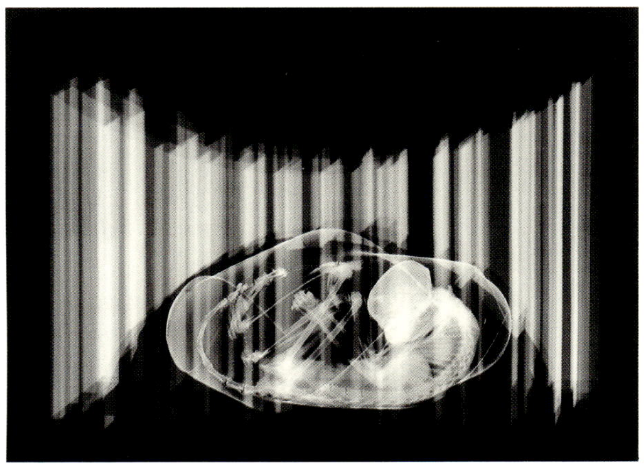

x=039 The fetus of a monkey, in its cosmic home that celebrates birth. In everthing, the end also marks the begining of the future. *Agachisu*; 1 twisted ellipsoid of vinyl chloride; 1 baby monkey.

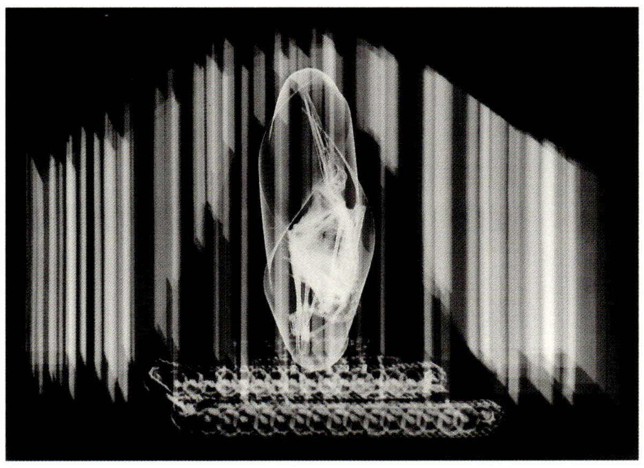

x=038 The artist's own grave. On the moving caterpillar tracks, a signature reads "Seiju" like an epitaph. The egg is a symbol of a return to life, a rebirth. Death can also signify life. *Agachisu* wood; 1 twisted ellipsoid of vinyl chloride; plastic toy-model tank's caterpillar tracks; lead lettering; 1 heron.

In compliance with all medical, legal, and moral considerations, the creatures that appear in this photo series are dead when used and are always treated with every respect. After their photographs are taken, the artist sees to their repose to the best of his ability, and returns them to the soil.

Seiju Toda: Career History

Stuart Klawans

Seiju Toda began his artistic career while a student at the public high school in Mikuni, the town in Fukui Prefecture where he had been born in 1948. Through a teacher at the school, Tadahiro Ono—an artist associated with Japan's avant-garde Ande Pada movement—Toda became acquainted with the work of the latest European and American artists. Of these, two in particular—John Cage and Jasper Johns—had a profound impact on the young student.

From Cage, Toda learned that an artwork could be changing and ephemeral—an event happening over time, rather than a static object. From Johns, he learned that the most ordinary things could be made into art, and that artworks could derive their meaning from the way different objects were assembled. Although such ideas were considered novel everywhere in the 1960s, they were especially striking in Japan, where artistic training placed a rigorous emphasis on technical skill.

Most shocking of all, perhaps, was the idea that it might be possible to deny traditional aesthetic values and yet affirm the value of art. "I saw," Toda has said, "that the avant-garde was not just a superficial style but a philosophy."

1965 : Performance Dedicated to John Cage, Mikuni High School.

I first saw the works of Jasper Johns and John Cage when I was in my sophomore year in high school. That was when I decided to pursue a career in art.

My first work, a performance put on at my high school, was dedicated to Cage. I ran two pieces of transparent plastic tubing onto the stage, one coming from the right, the other from the left. Offstage, I stationed volunteers. Their job was to drop marbles into the tubes. The marbles, which were covered in fluorescent paint, glowed as they rolled down the tubes, through a darkened space that was illuminated only by black light. When the marbles got to the center of the stage, they dropped from the ends of the tubes into two receptacles—one a wooden, Japanese-style washtub, the other a Western-style bucket made out of tin. That was my way of addressing the meeting of East and West, of paying tribute to John Cage, and of exploring chance effects of light and sound.

I like to come back to the origin of things. Artists such as Jasper Johns and Pablo Picasso overwhelm viewers with their talent. As for me, I hope to appeal to viewers with my heart.

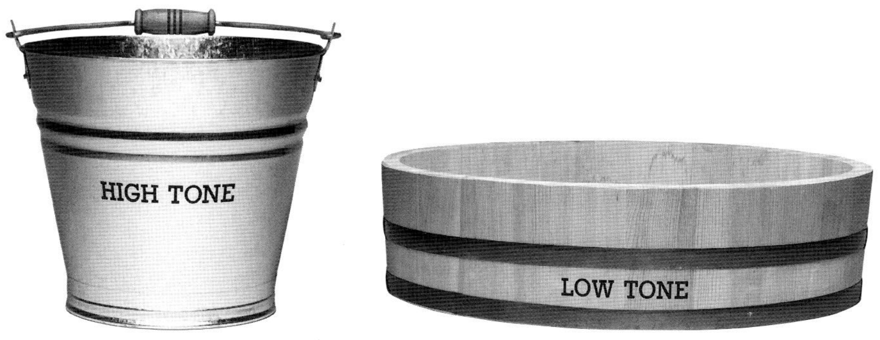

Chorus of Electric Guitar and Marbles.

1966 : *Yomiuri Independent* Exhibition, Sakai City, Osaka.

Each year, the *Yomiuri Independent* newspaper sponsored an exhibition featuring the works of prominent artists such as Shusaku Arakawa, On Kawahara, and Ushio Shinohara. In 1966, while still a high-school student, I found myself selected to exhibit in their company.

How could I make my presence felt among such artists? My answer was to create an installation titled "How to Get to the Place." I ran a piece of transparent plastic tubing through the entire exhibition, at the entrance to which I placed a tank filled with water that had been dyed blue. When visitors arrived, the first thing they saw was the blue water flowing into the tube. They could follow it all the way through the exhibition, past everybody else's work, to the place where the water flowed out—at my site.

The theme of "entrance" and "exit" still intrigues me, since it implies a kind of connection. By running water from a tank to the place where I stood, I created an entrance and an exit. In the same way, if I draw a line in white chalk from my studio to a museum, I am already creating and displaying a work of art. Or I could connect a line from my studio to a river—that, too, would represent "entrance" and "exit."

1967 : "Baby Exhibition" 「*Aka-chan Ten*」, Tokiwa Gallery, Tokyo.

My first chance to exhibit at an art gallery came in 1968, when the Tokiwa Gallery invited me to participate in a group show of emerging artists. For my contribution, I covered an entire wall of the gallery with objects made of droplets of transparent synthetic resin. The droplets resembled water. Through the installation, I wanted to express the importance of water to human life.

I also staged a performance at the same time. A bucket filled with water represented the human body. Droplets of water covered the outside of the bucket. That's the nature of water—it can fill things, or it can break down into beads. In my case, I broke the "water" into pieces with a hammer and then put them back into the container.

1968 : "Baby Exhibition," Tokyo.

For the 1968 "Baby Exhibition" at the Tokiwa Gallery, I confronted the theme of, "Which came first, the chicken or the egg?" I stacked 100 chicken coops in the gallery and put a white hen in each. Then, every time an egg was laid, I wrote the date on it and set it aside on display. During the one week of the exhibition, the hens laid 35 eggs. Eventually, of course, there were a lot more, since the installation was presented at various other galleries.

If I had to summarize the installation, I would say that it was trying to express the idea of the genetic code through the image of those neat ranks of hens and carefully dated eggs. Or you could put it another way and say it was about the transmigration of the soul.

1969 : Performance in Front of the Fukui Prefectural Office.

Even the most ordinary routine can come to have a new meaning, if it's done in a novel time and place. For example, you could drink your tea outdoors instead of in a restaurant, or sunbathe in a darkened room, or go through the motions of fishing indoors.

For this performance in front of the Fukui Prefectural Office, I put on a dark cloak, covered my head with bandages, and mimed eating a meal. (The main course, I recall, was soup.) Meanwhile, another performer in an identical costume silently mouthed a poem.

1969 : "Record Exhibition," PARCO Department Store, Tokyo.

In 1960 Jasper Johns made a bronze sculpture of two cans of Ballantine Ale. On one level, it was a negative gesture, since it denied the objects their function. You couldn't drink from Johns's ale cans. On another level, though, the gesture was positive, since it created art. You might say that Johns was "affirming the denial."

Can you take the process a step further? In 1967 I made sculptures out of some records—and the records could actually be played. They contained music by the Beatles and John Cage—and with their surfaces painted in red, blue, yellow, and green oil paint, they were more attractive than the originals.

Perhaps you could turn compact discs into works of art by marking them with the fingerprints of celebrities.

1969 : Three-man Exhibition "Baby Exhibition" 「*Aka-chan Ten*」, Tokiwa Gallery, Tokyo.

1970–73 : One-man Exhibition, Okabe Gallery and PARCO Department Store, Tokyo.

1974 : "The Eye of the Canvas and Light," Muramatsu Gallery, Tokyo.

My work with photography began in 1974 with this installation at Muramatsu Gallery. I mounted canvas panels around the walls of the darkened room. Each panel was printed with a photograph of a beam of light. A real beam of light, of exactly the same sort, was then focused on each panel.

When you closed your eyes and then opened them, the panel looked like either a painted canvas or a piece of sculpture. Either way, the image was created directly by light, which had become an integral part of the artwork.

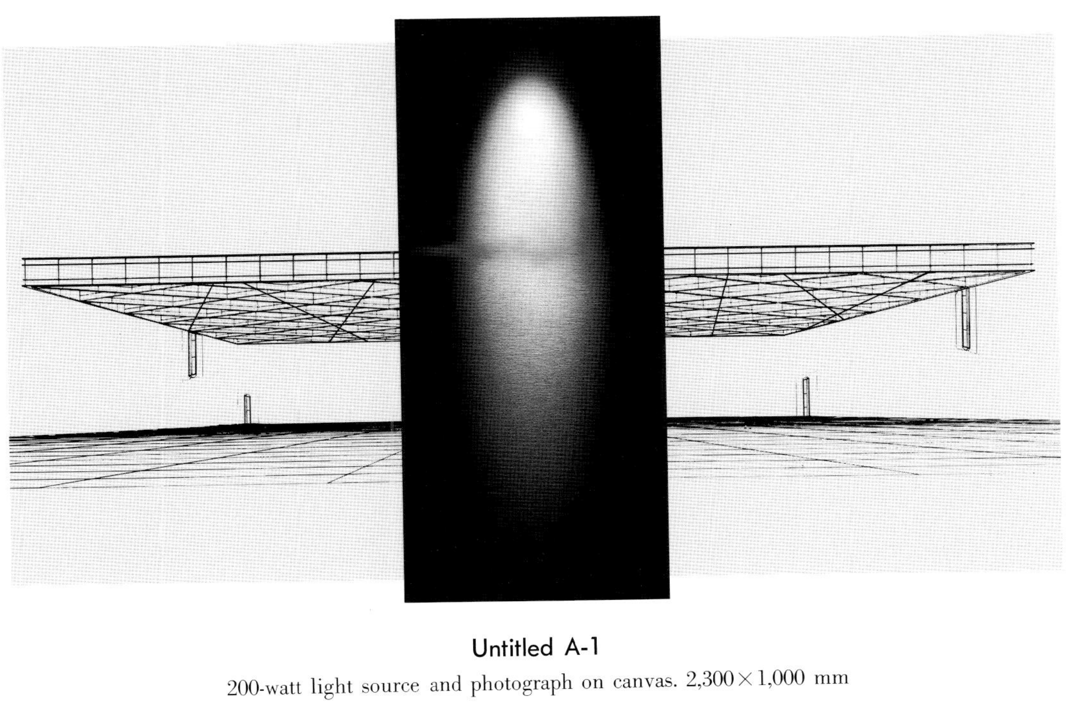

Untitled A-1

200-watt light source and photograph on canvas. 2,300×1,000 mm

1974 : One-man Exhibition, Muramatsu Gallery.

1974 : "Image '74 Group Exhibition," Okabe Gallery, Tokyo.

In the same year that I began working with photography, I also made my first exhibition of prints. In part, I was trying to extend my money-making work into the context of art—I had begun a career as a graphic designer. But I also felt that performance art and installation art were played out, at least for me. I had begun to want to make something that would last. The question was, how to make permanent art that would still express my concern with process?

For this exhibition, which also featured the work of Norio Imai and Kenshi, I presented a number of silkscreens titled *Japanese Craftsman's Series*. First, I took photographs of objects associated with different crafts—the carpenter's plank, the plasterer's tile. I made rough-grain prints of the photographs, then photocopied them, then based the silkscreen prints on the photocopies. The images themselves—the tile or the plank—were abstract and geometric. The methods of reproduction were mechanical. But the texture of the finished print made the viewer highly aware of the process of the art making, while also bringing to mind the handwork of the carpenter or plasterer.

1975 : 100-man Contemporary Block Print Exhibition. Mainichi Contemporary Art Exhibition. Tokyo Invitational Exhibition.

1975–76 : "Image '75 Group Exhibition," Okabe Gallery, Tokyo.

For this show, which included the work of Norio Imai, Masafumi Maita, and Kenshi, I again exhibited silkscreen prints based on photographs. This time, I took photographs of a piece of wood that was in motion, using multiple exposures so as to capture the movement. Then I printed the photographs as silkscreens, using metallic paint and black ink.

To me, movement itself is art. There are four kinds. First is movement in two dimensions—that is, on a plane. Then there is movement in three dimensions. The third type is movement in the imagination. The fourth type is movement with will. All four types may be forms of art.

1976 : One-man Exhibition, Fuma Gallery, Tokyo.

Once again, I made silkscreen prints of photocopies of photographs—this time basing the images on stacks of bricks, which I had wired together and then shattered. To create a second level of the installation, I also made silkscreens based directly on the negatives of the photographs. The *Asahi* newspaper, reviewing the exhibition, compared the work to "a poem . . . told through a naive bright shape, which is the physical metamorphosis of an image of delicate and pure charm."

In 1977 this series was shown at the Azuma Gallery, New York.

1977 : Miami International Block Print Biennial Exhibition. Bradford International Block Print Biennial Exhibition.
One-man Exhibition, Azuma Gallery, New York.

1978 : One-man Exhibition, Fuma Gallery, Tokyo.

In search of an alternative to painting on canvas, I had begun to think about a more direct and physically engaging way of image making: rubbings. You can make a painting of a sofa, say, or a desk, or an automobile. But you can also make a rubbing of the object's surface and get a much different result.

For this series, I prepared a concrete floor by treating it with synthetic resin, glass fiber, and oil colors. Then I drew on the floor: geometric designs, bands of color, series of lines, even images of ordinary objects, such as a bottle. Once I had made the drawings, I transferred those textured designs from the concrete floor onto another surface of synthetic resin. I liked the mixing of the colors, the varied textures, the collagelike effects.

Other works from this series were shown in 1978 at the Okabe Gallery, Tokyo.

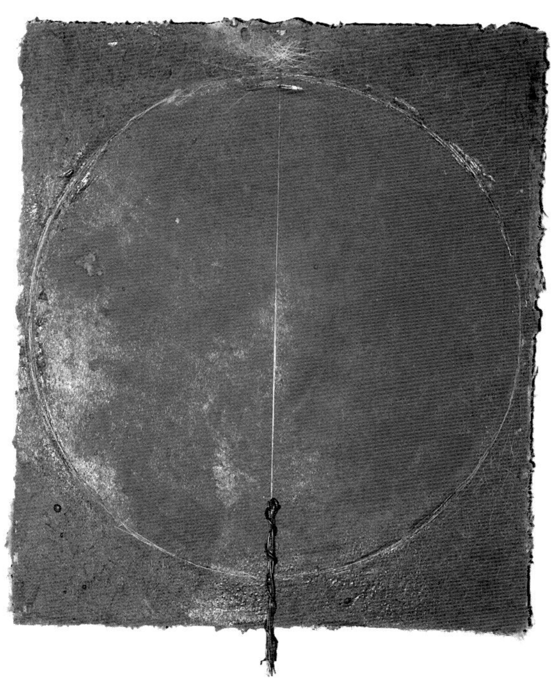

Bending Soup
synthetic resin, glass fiber, oil paints. 1,620×1,303 mm

1979 : Japan New Face Invitational Exhibition.

1980 : Miami International Block Print Biennial Exhibition.

25th Contemporary Block Print Exhibition, Tokyo Central Gallery.

CWAJ 25th Contemporary Block Print Exhibition, Tokyo American Club.

1981 : Canada International Block Print Biennial Exhibition.

Colorado International Invitational Poster Exhibition.

1981 : One-man Exhibition, Shirota Gallery, Tokyo.

Still searching for new surfaces on which to paint, I tried making a canvas that was not a canvas. Using glass fiber—with a mesh similar to that of coarse canvas—combined with synthetic resin and silicone, I made a molded, textured surface. Then, using oil paints, I added color, creating images that were geometric and abstract but highly textured and expressive in their brushwork.

Someday, I would like to make various "canvases" out of the transferred textures of roads from different countries around the world. I recently tried transferring textures from the ground onto different surfaces, and the results were unbelievably beautiful.

American Food

synthetic resin, glass fiber, oil paints. 1,620×1,303 mm

1981 : Hara Annual Invitational Exhibition, Hara Museum, Tokyo.

The chief element in this new surface is Japanese rice paper—*washi*—pasted onto canvas and then treated with silicone, synthetic resin, glass fiber, and oil paint. I wanted above all to emphasize the dry, powdery quality of the *washi.*

I feel that textures are more important than subject matter in my paintings. Sometimes I fantasize about creating a work on a large piece of cloth made from the skins of a whale and an elephant.

Untitled

synthetic resin, Japanese paper, powder of lead pencil. 7,000×10,000 mm

1982 : 20-man Poster Exhibition, Museum of Modern Art, Toyama.

1983 : One-man Exhibition, Shirota Gallery, Tokyo.

The theme of this exhibition was the Heian period (c. 800-1100), when Kyoto became the capital of Japan, feudal clans rose in power, and artists-such as Lady Murasaki in *The Tale of Genji*—reflected the concern of a refined aristocracy with the passing of earthly beauty. To me, the Heian period seems to be an era without shadows and without odors. To symbolize that, I made an installation of objects and prints based on plain Japanese cypress wood, hinoki.

I combined objects made of hinoki and other objects and photographed them on a cloudy day, using natural light, giving the pictures the shadowless effect I associate with the Heian period. All the objects were symbols of life, which is constantly changing. Hinoki changes, too—it warps and discolors within a day, as the temperature changes. To emphasize this theme, I put dates on each object. The photographs were taken at eye level—the angle that is least intimidating for the viewer—and the prints made from them were done in four colors with very little black.

For the installation, I covered the entire floor of the Shirota Gallery with tiles, to give the impression of a clean bathroom. Perhaps the most impressive object was a wooden ladder, on each rung of which, lying in red water, was a carp made out of plaster. A print of the same object was also in the exhibition, which the *Mainichi* newspaper called "strange and mysterious," "fresh and vivid," "a macabre image of the man-made world which surrounds us."

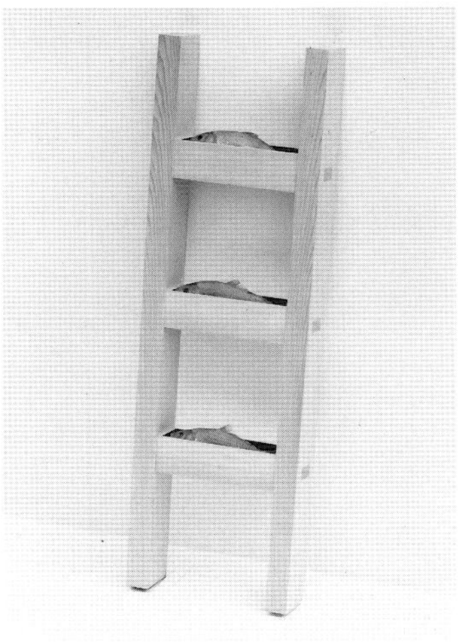

Red Whale

March Lion

Muscles Game

plain wood, carp, stingray, beef. 1,030×728 mm

Sweet Bible

Soft Bible

plain wood, rabbit, deer. 1,030×1,456 mm

1983 : Colorado International Invitational Poster Exhibition.

Ljubljana International Block Print Biennial.

1984 : Norway International Block Print Biennial.

Humor in Contemporary Art Exhibition, Museum of Modern Art, Saitama.

One-man Exhibition, Luft Gallery, Tokyo.

1985 : One-man Exhibition, Luft Gallery, Tokyo.

Pieces of plain wood are installed inside synthetic resin canvas frames.

Employing linear perspective to create solid objects have already been tried elsewhere; a new challenge rested in using the law of perspective to create flat surfaces. This seemingly impossible feat was accomplished by first stripping the drawings off their canvases and reassembling them into solid forms.

1985 : Photography and Print Project.

This group of photographs and prints represented the four seasons, each symbolized by a different plantlike form made of resin. The resin also resembled crystal. On the "plant" representing summer, I wrote "August 6, 1945"—the day the atomic bomb fell on Hiroshima.

Why did I want to make a "plant" out of crystal—in effect, out of light? I clearly remember the scene in *Close Encounters of the Third Kind* when the UFO, like a giant crystal, comes down to Earth, shining brightly. I thought it was very beautiful. I also remember being in a bar in the Ginza, talking with someone, when I happened to look up and see a chandelier hanging from the ceiling. The bright light of that chandelier is still printed in my mind.

And the brightest light in the world? The first people to look up and see it were in Hiroshima on that day in 1945.

First Atomic Bomb
crystal acryl, clay, salt, sand, ash. 9 pieces, 4,000×3,000 mm

1985 : One-man Exhibition, News Gallery, Tokyo.

I spread sand on the floor of the News Gallery and brought in three broken cars. In each of the cars was a video monitor. The image on the screen was the Happy Face character—only it did not remain happy. It would get angry, cry, laugh, then go back to its usual smile.

Human beings and machines alike share "the beauty of breakdown." That was the theme of this installation, which also was a way of crying out for peace. If I had put those Happy Face videos in another setting of breakdown and devastation—for example, an abandoned warehouse—they still would have made the point.

1987 : Installation of Issey Miyake Exhibition, Laforet Museum,Tokyo.

1988 : Invitation to the Ljubljana International Biennial Exhibition, Sakaide.

1989 : Budapest International Block Print Biennial Exhibition.
Invitation to the Ljubljana International Block Print Biennial Exhibition, Kawasaki City Museum, Japan.

1992 : Japanese Graphic Design Exhibition, Municipal Library of Berlin. Japanese Posters Exhibition, Museum für Gestaltung, Zurich.

1993 : Art History Exhibition, Tokyo Station Gallery. Japanese Posters Exhibition, Holland. Posters Festival, Chaumont, France.
Stop AIDS, ADC Posters Exhibition, Sendai. 80s Japanese Posters Exhibition, New Jersey. Design Shop Exhibition, Kyoto.

1994 : The 4th International Poster Triennial in Toyama Exhibition, The Museum of Modern Art, Toyama.
The World Design Exhibition, The Chicago Athenaeum. The 5th International Graphic Art Posters Festival, Chaumont, France.
Curator of Akira Kurosawa Drawings Exhibition, New York. The Contemporary Poster Exhibition "LIFE," Ginza Graphic Gallery.
Art Against AIDS Japan, "The Artist's Camera: Photographs by Contemporary Artists" Exhibition, Yokohama Landmark Hall and Tokyo Sogetu Art Gallery. The 100 Years Beautiful Woman Exhibition, Suntory Museum Tenpozan, Osaka.

Plus Many Other Invited Exhibitions.

Selected Major Awards

1979 : Special Prize at International Poster Biennial, Lahti.

1981 : Silver Award at '81 Japan-USA Poster Graphics Exhibition.

1983 : Highest Award of Excellence at International Poster Biennial, Lahti.

1984 : Silver Award at 4th International Poster Biennial, Toyama.

Mie Museum Prize at Contemporary Japanese Art Exhibition.

1984 : Gold Award, 6 Silver Awards, and 2 Bronze Awards at '84 Japan-USA Poster Graphics Exhibition.

1985 : Award for Excellence at the 1st Rublin International Anti-War Art Triennial.

1986 : Special Award at 11th International Poster Biennial, Warsaw.

1987 : Museum Award at 17th International Block Print Biennial, Ljubljana.

1989 : International Block Print Biennial Museum Award, Cracow.

1990 : Grand Prix at the International Design Exhibition of France.

1991 : Third Prize at International Poster Biennial, Lahti.

Collections

The Chicago Athenaeum

Colorado State University

The Denver Art Museum

Fukui Fine Arts Museum

Lahti Art Museum, Finland

Municipal Museum of Modern Art, Cracow

Municipal Museum of Munich, Germany

Museum für Kunst und Gewerbe of Hamburg, Germany

The Museum of Modern Art, New York

The Museum of Modern Art, Toyama

National Museum of Modern Art, Ljubljana

National Museum of Modern Art, Tokyo

National Museum of Oslo, Norway

The University of Connecticut

Warsaw Poster Museum

Staff

KUMIKO SHINOZUKA

TAKAMASA UOMOTO
MANABU OHASHI
MIKIRO ISHII
TAKESHI MASUMORI
AKINORI KIMURA
SHYOJI NISHIMORI

Special thanks to

Japan Industrial Testing Co., Ltd.
MASAHARU YASUKAWA

JUN MAKI
HIROSHI SUGIYAMA
YOSHINARI MATSUYAMA
MARIKO YOKOYAMA
TOHOKUSHINSHA FILM CORPORATION
IDEA OFFICE INC.

Toppan Printing Co., Ltd.
Plate Making: MITSUO OBOKATA
KIICHIRO TAMURA
SHINOBU SUZUKI
YASUHIKO YONEHARA
Printing Director: HIROSHI YOSHIDA
Printing and Book Binding: Toppan Printing Co.,Ltd.